Arising out of the author's lifetime fascination with the links between the formal language of mathematical models and natural language, this short book comprises five essays investigating both the economics of language and the language of economics. Ariel Rubinstein touches on the structure imposed on binary relations in daily language, the evolutionary development of the meaning of words, game-theoretical considerations of pragmatics, the language of economic agents, and the rhetoric of game theory. These short essays are full of challenging ideas for social scientists that should help to encourage a fundamental rethinking of many of the underlying assumptions in economic theory and game theory. A postscript contains comments by a logician, Johan van Benthem (University of Amsterdam, Institute for Logic, Language and Computation and Stanford University, Center for the Study of Language and Information) and two economists, Tilman Börgers (University College, London) and Barton Lipman (University of Wisconsin, Madison).

ARIEL RUBINSTEIN is Professor of Economics at Tel Aviv University and Princeton University. His recent publications include *Modeling Bounded Rationality* (1998), *A Course in Game Theory* (with M. Osborne, 1994) and *Bargaining and Markets* (with M. Osborne, 1990).

Economics and Language

Five Essays

THE CHURCHILL LECTURES IN ECONOMIC THEORY

Economics and Language

Five Essays

ARIEL RUBINSTEIN

Tel Aviv University
Princeton University

CAMBRIDGE
UNIVERSITY PRESS

PUBLISHED BY THE PRESS SYNDICATE OF THE UNIVERSITY OF CAMBRIDGE
The Pitt Building, Trumpington Street, Cambridge, United Kingdom

CAMBRIDGE UNIVERSITY PRESS
The Edinburgh Building, Cambridge CB2 2RU, UK www.cup.cam.ac.uk
40 West 20th Street, New York, NY 10011–4211, USA www.cup.org
10 Stamford Road, Oakleigh, Melbourne 3166, Australia
Ruiz de Alarcón 13, 28014 Madrid, Spain

First published 2000

Printed in the United Kingdom at the University Press, Cambridge

Typeface Trump Mediaeval 9.75/12 pt   System QuarkXPress® [SE]

A catalogue record for this book is available from the British Library

Library of Congress Cataloguing in Publication data
Rubinstein, Ariel.
Economics and language / Ariel Rubinstein.
p. cm.
Includes bibliographical references.
ISBN 0-521-59306-9 (hardbound)
1. Economics–Language. 2. Game theory. I. Title.
HB62.R83 2000
330′.01′4–dc21
ISBN 0 521 593069 hardback
ISBN 0 521 789907 paperback

CONTENTS

ACKNOWLEDGMENTS

This book emerged from the kind invitation of the fellows of Churchill College to deliver the Churchill Lectures of 1996. It is a pleasure for me to use this opportunity to thank Frank Hahn, for his encouragement during my entire career. The book is an extended version of my Churchill Lectures delivered in Cambridge, UK in May 1996. Part of the material overlaps with the Schwartz Lecture delivered at Northwestern University in May 1998.

Three people agreed on very short notice to provide their written comments and to publish them in the book: Johan van Benthem, Tilman Börgers and Bart Lipman. I thank them wholeheartedly for their willingness to do so. The idea to add comments to the book should be credited to Chris Harrison, the Cambridge University Press editor of this book.

Several people have commented on parts of the text in its various stages. In particular, I would like to thank Bart Lipman, Martin Osborne and Rani Spiegler, who were always ready to help, and Mark Voorneveld, who provided comments on the final draft of the manuscript. My special thanks to my partner in life, Yael. As so much else, the idea of the cover is hers.

My thanks to Nina Reshef and Sharon Simmer, who helped me with language editing; the remaining errors are my own.

Finally, I would like to acknowledge the financial support given me by the Israeli Science Foundation (grant no. 06-1011-0673). I am also grateful to The Russell Sage Foundation and New York University, which hosted me during the period when much of the writing was completed.

PART 1

ECONOMICS OF LANGUAGE

CHAPTER 0

ECONOMICS AND LANGUAGE

The psychologist Joel Davitz once wrote: "I suspect that most research in the social sciences has roots somewhere in the personal life of the researcher, though these roots are rarely reported in published papers" (Davitz, 1976). The first part of this statement definitely applies to this book. Though I am involved in several fields of economics and game theory, all my academic research has been motivated by my childhood desire to understand the way that people argue. In high school, I wanted to study logic, which I thought would be useful in political debates or in legal battles against evil once I fulfilled my dream of becoming a solicitor. Unfortunately, I became neither a lawyer nor a politician, and I have since come to understand that logic is not a very useful tool in these areas in any case. Nonetheless, I continued to explore formal models of game theory and economic theory, though not in the hope of predicting human behavior, not in anticipation of predicting the stock market prices, and without any illusion about the ability of capturing all of reality in one simple model. I am simply interested in the reasoning behind decision making and in the arguments people bring in debates. I am still puzzled, and even fascinated, by the magic of the links between the formal language of mathematical models and natural language. This brings me to the subject of this lecture – "Economics and Language."

0.1 Economics and language

The title of these lectures may be misleading. Although the caption "Economics and Language" is a catchy title, it is too vague. It encompasses numerous subjects, most of which will not be touched on here. This series of lectures will briefly address five issues which fall under this general heading. The issues can be presented in the form of five questions:

- Why do we tend to arrange things on a line and not in a circle?
- How is it that the utterance "be careful" is understood by the listener as a warning and not as an invitation to a dance?
- How is it that the statement "it is not raining very hard" is understood to mean "it is raining but not very hard"?
- Does the textbook utility function $\log(x_1 + 1)x_2$ make sense?
- Is the use of the word "strategy" in game theory rhetorical?

All the issues discussed in these lectures lie somewhere between economic theory and the study of language. Two questions spring to mind:

- *Why would economic theory be relevant to linguistic issues?* Economic theory is an attempt to explain regularities in human interaction and the most fundamental nonphysical regularity in human interaction is natural language. Economic theory carefully analyzes the design of social systems; language is, in part, a mechanism of communication. Economics attempts to explain social institutions as regularities deriving from the optimization of certain functions; this may be applicable to language as well. In these lectures I will try to demonstrate the relevance of economic thought to the study of language by presenting several "economic-like" analyses to address linguistic issues.

- *Why would economic theory be a relevant subject of research from the point of view of language?* Because economic agents are human beings for whom language is a central tool in the process of making decisions and forming judgments. And because the other important "players" in Economic Theory – namely ourselves, the economic theorists – use formal models but these are not simply mathematical models; their significance derives from their interpretation, which is expressed using daily language.

0.2 Outline of the lectures

The book deals with five independent issues organized into two groups:

Part 1 is entitled *"Economics of Language"* and comprises the core of this book. In Part 1, methods taken from economic theory are used to address questions regarding natural language. The basic approach is that language serves certain functions from which the properties of language are derived.

In *chapter 1*, I assume that language is the product of a "fictitious optimizer" who operates behind a "veil of ignorance." The substantive issue studied in this chapter is the structure imposed on binary relations in daily language. The designer chooses properties of binary relations that will serve the users of the language. The three parts of the chapter discuss three distinct targets of binary relations:

(1) To enable the user of the relation to point out nameless elements.
(2) To improve the accuracy with which the vocabulary spanned by the relation approximates the actual terms to which the user of the language is referring.
(3) To facilitate the description of the relation by means of examples.

It will be shown that optimization with respect to these three targets explains the popularity of linear orderings in natural language.

In *chapter 2*, we discuss the evolutionary development of the meaning of words. The analytical tool used is a variant of the game-theoretic notion of evolutionary stable strategy. Complexity considerations are added to the standard notion of evolutionary stable equilibrium as an additional evolutionary factor.

In *chapter 3*, I touch on pragmatics, the topic furthest from the traditional economic issues that are discussed in these essays. Pragmatics searches for rules that explain the difference in meaning between a statement made in a conversation and the same statement when it is stated in isolation. Grice examined such rules in the framework of a conversation in which the participants are assumed to be cooperative. Here, game-theoretical analysis will be used to explain a certain phenomenon found in debates.

Part 2 is entitled *Language of Economics* and includes two essays.

Chapter 4 deals with the *Language of Economic Agents*. The starting point of the discussion is that decision makers, when making deliberate choices, often verbalize their deliberations. This assumption is especially fitting when the "decision maker" is a collective but also has appeal when the decision maker is an individual. Tools of mathematical logic are used to formalize the assumption. The objective is to analyze the constraints on the set of preferences which arise from natural restrictions on the language used by the decision maker to verbalize his preferences. I demonstrate in two different contexts that the definability constraint severely restricts the set of admissible preferences.

Chapter 5 focuses on the rhetoric of game theory. Much has been written on the rhetoric of economics in general; little, however, has been written on the rhetoric of game theory. The starting point of the discussion is that an economic model is a combination of a formal model and its interpretation. Using the Nash bargaining solution as an illustration, I first make the obvious claim that differences in models which seem equivalent result in significant differences in the interpretation of their results. The main argument of the chapter is more controversial. I argue that the rhetoric of game theory is misleading in that it creates the impression that game theory is more "useful" than it actually is, and that a better interpretation would make game theory much less relevant than is usually claimed in the applied game theory literature.

Though the book covers several distinct issues under the heading of "economics and language," it by no means covers all the issues that might be subsumed under this rubric. For example, I do not discuss the (largely ignored) literature labeled the "economics of language" which was surveyed in a special issue of the *International Journal of the Sociology of Language* (see Grin, 1996). Grin (1996) defines the "economics of language" as "a paradigm of theoretical economics and uses the concepts and tools of economics in the study of relationships featuring linguistic variables; it focuses principally, but not exclusively, on those relationships in which economic variables play a part." This body of research does indeed revolve around traditional "economic variables" and related issues such as "the economic costs and benefits of multi-language society," "language-based inequality," and "language and nationalism." However, despite the similar headings, those issues are very far from my interests as expressed in this book.

0.3 One more personal comment

While browsing through the literature in preparation for these lectures, I came across a short article written by Jacob Marschack entitled the "The Economics of Language" (Marschak, 1965). The article begins with a discussion between engineers and psychologists regarding the design of the communications system of a small fighter plane. Following the discussion Marschak states: "The present writer ... apologizes to those of his fellow economists who might prefer to define their field more narrowly, and who would object to ... identification of economics with the search of optimality in fields extending beyond, though including, the production and distribution of marketable goods." He then continues: "Being ignorant of linguistics, he apologizes even more humbly to those linguists who would scorn the designation of a simple dial-and-buttons systems a language." I don't feel that any apology is due to economists ... but I do feel a sincere apology is owed to linguists and philosophers of language. Although I am quite ignorant in those areas, I hope that these essays present some interesting ideas for the study of language.

REFERENCES

Davitz, J. (1976) *The Communication of Emotional Meaning,* New York: Greenwood Publishing Group

Grin, F. (1996) "Economic Approaches to Language and Language Planning: An Introduction," *International Journal of the Sociology of Language,* 121, 1–16

Marschak, J. (1965) "The Economics of Language," *Behavioral Science,* 10, 135–40

CHAPTER 1

CHOOSING THE SEMANTIC PROPERTIES OF LANGUAGE

1.1 Introduction

This chapter will present a research agenda whose prime objective is to explain how features of natural language are consistent with the optimization of certain "reasonable" target functions. Rather than discuss the research agenda in abstract, I will begin with the specific argument and return to the general discussion at the end of the chapter.

This chapter discusses binary relations. A binary relation on a set Ω specifies a connection between elements within the set. Such binary relations are common in natural language. For example, "person x knows person y," "tree x is to the right of tree y," "picture x is similar to picture y," "chair x and chair y are the same color," and so on. I will avoid binary relations such as "Professor x works for university y" or "the Social Security number of x is y," which specify "relationships" between elements which naturally belong to two distinct sets. I will further restrict the term "binary relation" to be irreflexive: No element relates to itself. The reason for this is that the term "x relates to y" when $x=y$ is fundamentally different from "x relates to y" when $x \neq y$. For example, the statement "a loves b" is different from the statement "a loves himself."

Certain binary relations, by their nature, must satisfy certain properties. For example, the relation "x is a neighbor of y" must, in any acceptable use of this relation, satisfy the symmetry property (if x is a neighbor of y, then y is a neighbor of x). The relation "x is to the right of y"

This chapter is based on Rubinstein (1996).

9

must be a *linear ordering*, thus satisfying the properties of completeness (for every $x \neq y$, either x relates to y or y to x), asymmetry (for every x and y, if x relates to y, y does not relate to x), and transitivity (for every x, y, and z, if x relates to y and y to z, then x relates to z). In contrast, the nature of many other binary relations, such as the relation "x loves y," does not imply any specific properties that the relation must satisfy *a priori*. It may be true that among a particular group of people, "x loves y" implies "y loves x." However, there is nothing in our understanding of the relation "x loves y" which necessitates this symmetry.

The subject of this chapter is in fact the *properties* of those binary relations which appear in natural language. Formally, a property of the relation R is defined to be a sentence in the language of the calculus of predicates which uses a name for the binary relation R, variable names, connectives, and qualifiers, but does not include any individual names from the set of objects Ω. I will refer to the combination of properties of a term as its *structure*.

I am curious as to the structures of binary relations in natural language. I search for explanations as to why, out of an infinite number of potential properties, we find that only a few are common in natural languages. For example, it is difficult to find natural properties of binary relations such as the following:

A1: If xRy and xRz ($y \neq z$), and both yRa and zRa, then also xRa.

A2: For every x there are *three* elements y for which xRy. (In contrast, the relation "x is the child of y" on the set of human beings does satisfy the property that for every x there are *two* elements y which x relates to.)

Alternatively, it is difficult to find examples of natural structures of binary relations which are required to be tournaments (satisfying completeness and asymmetry) but which are not required to satisfy transitivity. One exception is the structure of the relation "x is located clockwise from y (on the shortest arc connecting x and y)." Is it simply a coincidence that only a few structures exist in natural language?

The starting point for the following discussion is that binary relations fulfill certain functions in everyday life. There are many possible criteria for examining the functionality of binary relations. In this discussion, I examine only three. I will argue that certain properties, shared by linear orderings, perform better according to each of these criteria. Of course, other criteria are also likely to provide alternative explanations for the frequent use of various common structures such as equivalence and similarity relations.

1.2 Indication-friendliness

Consider the case in which two parties observe a group of trees and the speaker wishes to refer to a certain tree. If the tree is the only olive tree in the grove, the speaker should simply use the term "the olive tree." If there is no mutually recognized name for the tree and the two parties have a certain binary relation defined on the set of trees in their mutual vocabulary, the user can use this relation to define the element. For example, the phrase "the third tree on the right" indicates one tree out of many by using the linear ordering "x stands to the left of y" when the group of trees is well defined and the relation "being to the left of" is a linear ordering. Similarly, the phrase "the seventh floor" indicates a location in a building given the linear ordering "floor x is above floor y." There would be no need to use the phrase if it was known to be "the presidential floor." On the other hand, the relation "line x on the clock is clockwise to line y (with the smallest angle possible)" does not enable the user to indicate a certain line on a number-less clock; any formula which is satisfied by three o'clock is satisfied by four o'clock as well. In fact, the existence of even one designated line such as "twelve o'clock", would enable the use of the relation to specify all lines on the clock. The effect of using such a designated element is equivalent to transforming the circle into a line.

Thus, binary relations are viewed here as tools for indicating elements in a set whose objects do not have names.

We look for structures that enable the user to unambiguously single out *any element out of any subset of Ω*. We are led to the following definition:

Definition: A binary relation R on a set Ω is *indication-friendly* if for every $A \subseteq \Omega$, and every element $a \in A$, there is a formula $f_{a,A}(x)$ (in the language of the calculus of predicates with one binary relation and without individual constants) such that a is the only element in A satisfying the formula (when substituting a in place of the free variable x).

All linear orderings are indication-friendly. If R is a linear ordering, the formula $P_1(x) = \forall y(x \neq y \rightarrow xRy)$ defines the "maximal" element in the set A for $A \subseteq \Omega$. The formula $P_2(x) = \forall y(x \neq y \wedge -P_1(y) \rightarrow xRy)$ defines the "second-to-the-maximal" element, and so on. Note that in natural language there are "short cuts" for describing the various elements. For example, the "short cuts" for $P_1(x)$ and $P_2(x)$ are "the first" and "the second."

In contrast, consider the set $\Omega = \{a,b,c,d\}$ and the non-linear binary relation R, called "beat," depicted in the following diagram (aRb, aRc, dRa, bRd, bRc and cRd):

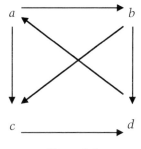

Figure 1.1

Referring to the grand set Ω, the element a is defined by "it beats two elements, one of which also beats two elements." The element b is defined by "it beats two elements, which each beat one element." And so on. However, whereas the relation R allows the user to define any element in the set Ω, the relation is not effective in defining elements in the subset $\{a,b,d\}$, in which case the induced relation is cyclical.

We will now demonstrate that if Ω is a finite set and R is a binary relation, then R is indication-friendly if and only if R is a linear ordering. We have already noted that if R is a linear ordering on Ω, then for every $A \subseteq \Omega$ and every $a \in A$, there is a formula which indicates a. Assume that a binary relation R is indication-friendly. For any two elements $a,b \in \Omega$, in order to indicate any of the elements in the two-member set $A = \{a,b\}$, it must be that either aRb or bRa but not both; thus, R must be complete and asymmetric. R must *also* be transitive since for every three elements $a,b,c \in \Omega$, the relation must not be cyclical in order to indicate each of the elements in the set $A = \{a,b,c\}$.

Conclusion 1: A binary relation enables the user to indicate any element in any subset of the grand set if and only if it is a linear ordering. Linear orderings are the most efficient binary relations for indicating every element in every subset.

1.3 A detour: splitting a set

We will now make a brief detour from the world of binary relations in order to discuss unary relations. Assume that the user of the language can refer to a set of objects X (such as "the set of flowers"). From time to time, the speaker will wish to refer to subsets of X (either for the purpose of conversing with another person or for storing information in his own mind). However, he will be able to refer only to terms that appear in his language. Initially, the speaker can refer to the set of all Xs ("pick all flowers") or to the null set ("don't pick any flowers"). In order to extend his vocabulary the designer of the language is permitted to invent (given "hardware" constraints) one additional term for one subset of X. The objective of the designer is to introduce one new term so that the speaker can refer to a new set in as accurate a manner as possible (on average).

Let us be more precise: Restrict X to be a finite set. If the term "the set S" is well defined, then the user will have four expressions available for referring to subsets of X: "all Xs," "the Ss," "the not Ss," and "nothing." We will call the collection of sets $V(S) = \{X, S, -S, \emptyset\}$ the *vocabulary* spanned by S.

The basic idea is that language should be flexible enough to function under unforeseen circumstances. The speaker can use the terms in $V(S)$ but will eventually need to refer to sets not necessarily contained in $V(S)$. The term S will be evaluated by the vocabulary's ease of use with "least loss." It is assumed that when the speaker wishes to refer to a set $Z \subseteq X$, he will use an element in $V(S)$ which is "closest" to the set Z. In order to formally state this idea, we need to define the *distance* between two sets. Let the distance between A and B, $d(A,B)$ be the cardinality of the asymmetric difference between A and B (the set of all elements which are in A and not in B or in B and not in A) – i.e. $d(A,B) = |(A-B) \cup (B-A)|$. One interpretation of this distance function fits the case in which the user who wishes to refer to the set B and employs a term $A \in V(S)$ lists the elements in B which are excluded from A or appended to A and bears a "cost" measured by the number of elements which have to be excluded or appended (for example, the sentence "You may eat only bread or any fruit with the exception of apples and bananas" utilizes the term fruit and three individual names: bread, apples, and bananas).

For a set B and a vocabulary V, define $\delta(B,V) = \min_{A \in V} d(A,B)$, the distance of the set B from the closest set in the vocabulary V. By assigning equal "probabilities" *a priori* to all possible sets that the user might wish to refer to, the problem for the designer becomes $\min_S \Sigma_{Z \subseteq X} \delta(Z, V(S))/2^{|X|}$. Essentially, the problem boils down to the choice of the *number* of elements in the optimal set S.

To *further* clarify the nature of the problem, we will carry out a detailed calculation for a four-element set X. Consider the case in which S contains two elements. The user can refer to any of the sets X, $\emptyset$, S or $-S$, without incurring any loss. He can approximate any one- or three-element set by using the set ϕ or the set X, respectively, while bearing a "cost" of 1. If he wishes to refer to one of the four two-element sets which are not S or $-S$, he incurs a loss of 2. Thus, the average loss is $[4(0)+8(1)+4(2)]/16 = 1$. A similar calculation leads to the conclusion that a choice of S as $\emptyset$ or X leads to an expected loss of 5/4 (no loss for $\emptyset$ and X, a loss of 1 for the eight sets of size 1 and 3, and a loss of 2 for the six sets of size 2 – i.e. $[2(0)+8(1)+6(2)]/16 = 5/4$). If S is taken to be a one-element set (or a three-element set), the average loss is only $[4(0)+6(1)+6(1)]/16 = 3/4$. Thus, splitting X into two *unequal* subsets of sizes 1 and 3 minimizes imprecision.

Despite the above results, our intuition is correct in that the optimal size of S should be half that of X. When the set X is "large," the loss associated with choosing an S containing half of X's members is "close" to being minimal. The following proposition is an exact statement of this result (its proof, omitted here, includes a combinatorial calculation):

Claim: Let X be an n-element set. The difference between the solution of the problem $\min_S \sum_{z \subseteq X} \delta(Z, V(S))/2^{|X|}$ and the expected loss from the optimal use of the vocabulary spanned by an $(n/2)$-member subset of X, is in the magnitude of $1/n^{1/2}$.

1.4 Informativeness

We now return to the world of binary relations on some set Ω. An additional function of binary relations on a set Ω is to transfer or store information concerning a specific relationship existing between the elements of Ω. Consider the case in which the grand set includes all authors of articles in some field of research and the speaker is interested in describing the relation "x quotes y in his article." The speaker may describe the relation by listing the pairs of authors who satisfy the relation. Alternatively, he may use those binary relations which are available in his vocabulary to describe the "x quotes y" relation. If he finds his vocabulary insufficient to describe the relation, he will use a binary relation which is the best approximation. For example, if the relation "x is older than y" is well defined, the speaker can use the sentence: "An author quotes another if he is older than himself." As this may not be entirely correct, he can add a qualifying statement such as "the exceptions are a who did not quote b (though b is older), and c who did quote d (though d is not older)." Such qualifying statements are the "loss" incurred from the use of an imprecise relation in order to approximate the "who quotes whom" relation.

Our discussion envisages an imaginary "planner" who is able to design only one binary relation during the "initial stage of the world." Of course, real-life language includes numerous relations and the effectiveness of each depends on the entire fabric of the language. The assumption that the designer is planning only one binary relation is made solely for analytical convenience.

The design of one binary relation allows the speaker to select one of four binary relations. For instance, he can state: "Every author is quoted by all others younger than him," or "Every author is quoted by all others not younger than him," and of course he can also state "Everyone quotes everyone" and "No one quotes anyone," which do not require familiarity with any binary relations. Given a relation R, we will refer to these four relations as the vocabulary spanned by R and denote it by $V(R)$. (Note that in defining the vocabulary spanned by R, we ignore other possibilities for defining a binary relation using R, such as statements of the type "xSy if there is a z such that xRz and zRy.")

We assume that the speaker who wishes to refer to a binary relation S will use a relation in $V(R)$ which is the "closest" approximation. The loss is measured by the number of differences between the relation which the speaker wishes to describe and the one he finds available in his vocabulary. The distance between any two binary relations R' and R'' is taken to be the number of pairs (a,b) for which it is not true that $aR'b$ if and only if $aR''b$. Note that according to this measure, any pair for which R' and R'' disagree receives the same weight. Regarding the initial state, it seems proper to put equal weights on all possible "imprecisions." The designer's problem is to minimize the expected loss resulting from the optimal use of his vocabulary. It is assumed that from his point of view, all possible binary relations are equally likely to be required by the speaker. Thus, the designer's problem is $\min_R \sum_S \delta(S, V(R))$, where $\delta(S, V(R)) = \min_{T \in V(R)} \delta(S, T)$ and $\delta(S, T)$ is the distance between the relations S and T.

Stated in this way, the problem becomes a special case of the problem discussed in the previous section. The set X, in the terminology employed in that section, is the set $\Omega \times \Omega - \{(\omega, \omega) | \omega \in \Omega\}$. A binary relation R is identified with the graph of R – i.e. the subset of X of all (a, b) for which aRb. Recall that we concluded earlier that splitting the set X into two equal sets is nearly optimal for the designer if he wishes to reduce the expected number of "imprecisions."

We have one more step to go to reach the goal of this section. In planning a binary relation on Ω, the designer may also consider the possibility that the relation will eventually be used in reference to a subset of Ω. This is analogous to a binary relation being indication-friendly if it allows the indication of any object out of *any subset of objects* (see section 1.2). Hence, R has to be "optimal" for potential use in indicating a subset of $\Omega' \times \Omega' - \{(\omega, \omega) | \omega \in \Omega'\}$ for every subset $\Omega' \subseteq \Omega$. If R is complete and asymmetric, for every subset $\Omega' \subseteq \Omega$, the induced relation $R_{|\Omega'}$ (defined by $aR_{|\Omega'}b$ iff $a, b \in \Omega'$ and aRb) includes exactly half the pairs in $\Omega' \times \Omega' - \{(\omega, \omega) | \omega \in \Omega'\}$.

In summary, our fictitious planner wishes to design a binary relation which spans a vocabulary with the goal of minimizing the expected inaccuracy of the term the user will actually use. Viewing a relation as a set of pairs of elements, the problem was linked to the optimization problem discussed in the previous section. There we concluded that splitting a set into two subsets allows "close to optimal" use of the induced vocabulary. Requiring the relation to be complete and asymmetric guarantees that for any subset of the grand set, the restricted relation will be "close to optimal" as an aid to the user in specifying a relation on the subset.

Conclusion 2: In order to express binary relations as accurately as possible on any subset of a set Ω using a vocabulary spanned by a single binary relation on the set

Ω, a binary relation on Ω which is complete and asymmetric is close to optimal.

1.5 Ease of describability

In this section we discuss the third and last criterion by which binary relations are assessed in this chapter. Imagine that a hunter wishes to instruct his son on how to behave in the forest. When he observes two potential animals a and b, should he pursue a or b? The instructions have to be applicable to any pair of animals and must be clear. Thus, the set of instructions can be represented by a complete and asymmetric binary relation R where aRb means that when the son simultaneously observes a and b he should pursue a.

The son is aware of the sets of animals that are edible and the structure of the relation R (i.e., the list of properties satisfied). The son is acquainted with the structure of the relation either because it is instinctual or because his father has informed him of these properties. Therefore, all that is left for the father to do, when transferring the content of R to his son, is to provide him with a list of "examples" – i.e., statements of the type "animal a should be pursued when you see it together with animal b." The examples should be rich enough to allow the son to infer the entire relation from the structure and examples.

To illustrate, consider the case in which the relation R is a linear ordering and the number of elements in Ω is n. The minimal number of examples that the father must provide in this case is $n-1$ ($a_1 R a_2, a_2 R a_3 \ldots, a_{n-1} R a_n$).

This brings us to the main topic of our discussion. We assume that providing examples is costly. (The complexities of the structure and process of making the inferences are ignored here.) The following problem comes to mind: What are the structures of the complete and asymetric binary relations (tournaments) for which the number of examples required for their description is minimal?

Definitions: We say that $(f, \{a_i R b_i\}_{i \in I})$ defines the binary relation R^* on Ω when

- f is a sentence in the language of the calculus of predicates with one binary relation named a_i, b_iR and for all i, $a_i, b_i \in \Omega$
- R^* is the unique binary relation on Ω satisfying the sentence f and for all i, it is true that $a_iR^*b_i$.

The *complexity* of R^*, denoted $\ell(R^*)$, is the minimum number of examples that needs to be appended to R^* in order to enable a definition of R^* – i.e., the minimum size of the set I on all possible definitions of R^*.

As previously stated, our attention is limited to binary relations which are tournaments. The mathematical problem we wish to solve is:

$$\min_{R^* \text{ is a tournament}} \ell(R^*).$$

Note that for a given (finite) set of objects $\Omega = \{a_1,...,a_n\}$, the "structure" of any binary relation R^* can be expressed by the sentence $\varphi_{R^*}(x_1, ..., x_n) = \exists x_1, ..., x_n (\bigwedge \{x_iRx_j | a_iR^*a_j\})$. Thus, the optimization problem proposed above is equivalent to the following "puzzle-type" problem: Start with the graph of a tournament in which the names of the vertices have been erased. What is the minimum number of examples required to recover the names of the vertices (up to isomorphism)?

Example 1: Consider the tournament R^* on the set $\Omega = \{a,b,c\}$ where aR^*b, bR^*c and cR^*a. The sentence which states that R is complete, asymmetric, and anti-transitive $(\forall x,y,z(xRy \text{ and } yRz \to -xRz))$ is consistent with two relations on Ω; *hence*, a single observation, aRb, is needed to complete the definition of R^*. Thus, $\ell(R^*) = 1$.

Example 2: Let R^* be a linear relation. The relation is defined by a sentence f expressing completeness, asymmetry and transitivity, and by $n-1$ examples $\{a_iR^*a_{i+1}\}_{i=1,...,n-1}$, where $a_1R^*a_2R^*a_3, ..., a_{n-1}R^*a_n$. Obviously, there is no definition of a linear relation with less than $n-1$ observations. Thus, $\ell(R^*) = n-1$.

Example 3: Let $\Omega = \{a,b,c,d\}$ $(n=4)$ and let R^* be the relation satisfying aRx for all x and $bRcRdRb$. R^* is defined by

$\exists wxyz[wRx \wedge wRy \wedge wRz \wedge xRy \wedge yRz \wedge zRx]$ and the three observations aRb, aRc and bRc.

Example 4: Consider the relation R^* on the set $\Omega = \{a,b,c,d\}$ described in figure 1.2. The structure of the relation R can be formulated by a sentence of the type $\exists v_1 v_2 v_3 v_4 \varphi(v_1, v_2, v_3, v_4)$. Twenty-four different binary relations on Ω have this structure. It is simple to verify that $\ell(R^*) = 4$.

Example 5 (Fishburn, Kim and Tetali, 1994):

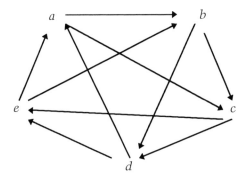

Figure 1.2

This relation, R^*, defined on the 5-element set $\Omega = \{a,b,c,d,e\}$ satisfies $\ell(R^*) = 3$. It is nicely defined by a sentence expressing the property that for every x there are precisely two elements "beaten" by x. The three observations aRb, aRc, and eRb, define the relation through the chain of conclusions $\{dRa, eRa\}$, $\{cRe, dRe\}$, $\{bRd, cRd\}$ and, finally, bRc.

Examples 2 and 5 illustrate that for $n = 3$ and $n = 5$, a linear ordering is not the most "economical" structure. Are there any other binary relations with $n > 7$ which are defined by less than $n - 1$ observations? I am not aware of a complete answer to this question. However, we do know the following (this proposition was suggested to me by Noga Alon):

Proposition 1.3: For any ϵ there exists $n(\epsilon)$ such that for any $n > n(\epsilon)$ and for any complete and asymmetric relation R on a set of n elements, $\ell(R) > (1 - \epsilon)n$.

Thus, at least for large sets, linear orderings are "almost" optimal with respect to the criterion of minimizing the number of observations required for their definition.

Comment: Notice that in the above discussion, we allowed the relation to be defined by a formula which depends on the number of elements in the set Ω. In contrast, the properties of linear orderings are expressed by a formula which does not depend on the number of elements in the set Ω. This leads to the following conjecture:

Conjecture: Let φ be a sentence in the language of the calculus of predicates which includes a single name of a binary relation R. There exists n^* such that if $|\Omega| \geq n^*$; then, for any tournament R^* which is defined on the set Ω by the sentence φ, $\ell(R^*) \geq |\Omega| - 1$.

In other words, although one can define a relation for "small" sets with less than $|\Omega| - 1$ examples, it is conjectured that a sentence must be accompanied by at least $|\Omega| - 1$ elements in order to define a tournament when the size of the set is "large enough."

Comment: We conclude this section with an explanation of why the term "describability" – and not "learnability" – is used in this section. In our scenario the father chooses the examples he presents to his son. The number $\ell(R^*)$ is the minimum number of examples the father has to present in order to convey enough information for the son to deduce the content of R^*. When choosing the examples, the father knows the relation which he would like his son to learn. On the other hand, had the son wished to acquire the content of the relation R^* by asking a list of "questions" of the type "what should be chased, a or b?," he would not necessarily have asked first for the $\ell(R^*)$ "examples" which could convey the relation R^*. For example, he would need 2.5 inquiries "on average" in order to infer a linear ordering defined on a three-element set.

1.6 Discussion

This chapter investigated the observation that in a natural language, certain structures of binary relations appear much more frequently than others; in particular, we discussed properties that are satisfied by linear orderings. It was argued that certain functions of binary relations in natural languages are better served by relations satisfying these properties. I believe that this is no more than an interesting fact. A stronger interpretation of the results requires establishing a connection between the optimality considerations presented above and the realization of the optimal solutions in the real world. Such a connection would require at least three premises:

(1) In order to function, natural languages include only a small number of structured binary relations.

(2) Binary relations fulfill several functions in natural languages.

(3) There are forces (evolution or a planner) which make it more likely that structures which are "optimal" with regard to the functions of binary relations will be observed in natural languages.

The first premise states that language inherently exhibits few of the properties of binary relations. Only if the number of possible structures is small can a user of the language deduce a relation's structure from a small number of instances in which the relation is used. (It is amazing how few observations of the type "*a* is better than *b*" are sufficient to teach a child that this relation is transitive.)

The second premise is the central one in this discussion. There are numerous potential criteria by which to measure the functionality of binary relations. Three such criteria were examined above. It was argued that certain properties (all shared by linear orderings) perform better according to each of these criteria.

The third premise, which links the first two, states that either there is a linguistic "engineer" who chooses the properties of binary relations so that they function effectively or that evolutionary forces select structures which are optimal or nearly so with respect to the functions they fulfill. This idea, which is popular in economics, has also been noted by philosophers. For example, Quine states: "If people's innate spacing of qualities is a gene-linked trait, then the spacing that has made for the most successful inductions will have tended to predominate through natural selection" (Quine, 1969, p.126).

The approach adopted in this chapter is related to the functionality of the language approach discussed in linguistics (see, for example, Piatelli-Palmarini, 1970) and to attempts to explain the classification system in natural language (see Rosch and Lloyd, 1978).

The discussion in this chapter is also related to classical philosophical discussions on "natural kinds." The notion of a "natural kind" emerges from the philosophical inquiry into the factors which confirm an inductive argument (see, for example, Goodman, 1972, Quine, 1969, Watanabe, 1969. A key puzzle in this literature is the so called "riddle of induction": Let us say that up to this moment, all observed emeralds were green. Why does this observation imply that all emeralds are green rather than all emeralds are "grue," an alternative category which includes all objects which were green up to this moment and blue from now on?

One possible answer to this question is that the inductive process relies on notions of similarity. Inductive arguments are made only with regard to categories of similar objects. The category green contains similar elements; grue does not. A green element yesterday and a green element tomorrow are similar; in contrast, a grue element yesterday is not similar to a grue element tomorrow. However, this only begs the question since one is left with the problem of determining the natural similarity relations. Here we run into similar difficulties. One possible solution is to argue that two objects are similar if "most" unary predicates coincide in satisfying the two objects. The difficulty with this criterion is raised by "The Ugly Duckling Theorem" (see Watanabe, 1969, Section 7.6). If the set of predicates is closed under Boolean operations, then the number of predicates which satisfies any possible object is constant; thus, the existence of elementary unary predicates cannot be the basis for explaining the existence of specific similarity relations. This led Watanabe to conclude that "if we acknowledge the empirical existence of classes of similar objects, it means that we are attaching nonuniform importance to various predicates, and that this weighting has an extra-logical origin" (Watanabe, 1969, p. 376). Thus, there is no escape from assuming that a certain kind of predicate (like "green" and not like "grue") has a preferred status called "natural kind" (see Quine, 1969).

Within the class of properties of binary relations, linear orderings, more than other structures, appear to be of a "natural kind." This chapter has attempted to provide some rationale as to why this is so.

REFERENCES

Fishburn, P., J.H. Kim and P. Tetali (1994) "Tournament Certificates," AT&T Bell Labs, mimeo
Goodman, N. (1972) *Problems and Projects*, Indianapolis and New York: Bobbs-Merrill
Piattelli-Palmarini, M. (ed.) (1970) *Language and Learning: The Debate between Jean Piaget and Noam Chomsky*, Cambridge, MA: Harvard University Press
Quine, W.V. (1969), *Ontological Relativity and Other Essays*, New York: Columbia University Press
Rosch, E. and B. Lloyd (eds.) (1978) *Cognition and Categorization*, Hillsdale: Lawrence Erlbaum Associates
Rubinstein, A. (1996) "Why are Certain Properties of Binary Relations Relatively More Common in Natural Language?," *Econometrica*, 64, 343-56
Watanabe, S. (1969), *Knowing and Guessing*, New York: John Wiley

CHAPTER 2

EVOLUTION GIVES MEANING TO LANGUAGE

2.1 The lack of meaning of words in games

Let us imagine a community of fishermen who make daily fishing trips onto the lake in small two-man boats. Occasionally, one fisherman notices a rock in the water which requires his partner to take evasive action. In such a situation, the observer needs to warn his partner. It would appear that the observer has only to shout "Be careful" in order to transmit the information about the obstacle. But why is "Be careful" indeed interpreted by his partner in the way we understand the phrase rather than as "Here is a fish" or as "What a beautiful day it is"? This is the question addressed in this chapter.

The situation is an interactive one in which the meaning of a statement depends on what the speaker thinks that the listener understands, which in turn depends, in a circular way, on what the listener thinks that the speaker has in mind. Thus, the situation can be analyzed in a game-theoretic framework. The basic situation will be viewed as a symmetric game. A player can either be an observer or a listener. A strategy is a plan of action for reacting to an

The chapter is based on variations of Maynard Smith's concept of an ESS (see Maynard Smith, 1982). The application of these ideas to the evolution of communication protocols was explored by Blume, Kim and Sobel (1993) and Warneryd (1993), among others. The ideas in this chapter are closely connected to the work of Banerjee and Weibull (1993) and Warneryd (1998), who recognized the power of using "complexity considerations" in the evolutionary analysis of "language formation." For a formal introduction of complexity considerations into game situations, see Rubinstein (1986). For studies of complexity considerations within an evolutionary framework, see Binmore and Samuelson (1992) and Binmore, Piccione and Samuelson (1998).

obstacle in front of the boat and responding to the partner's cry while rowing the boat. Note (at this stage informally) that the situation has one "equilibrium" in which the cry is made in the case of an obstacle being spotted and is interpreted correctly. However, the situation has another "pooling" equilibrium in which an observer ignores the rock (remains silent) and is not "programmed" to respond to his partner's shout if there is one.

The multiplicity of equilibria in this situation is one of the most "frustrating" challenges encountered by game theoreticians. In fact, it is difficult to describe communication using game-theoretic models owing to a fundamental property of any game-theory model: The familiar solution concepts are "invariant" to the names of the actions. Consider a game and rename the actions: Whatever game-theoretic solution concept was used for the first game is transformed into an equilibrium of the game with the actions renamed!

One exception is Farrell's solution concepts, which are related to "cheap talk" (see, for example, Farrell, 1993). Farrell assumes that words have a *fixed* external meaning. A listener operates under the assumption that unless a speaker has a "good reason" to mislead him, he does not do so. Farrell argues for the instability of the pooling equilibrium in which an observer remains silent and a listener ignores the cry "Be careful." This equilibrium will be "destroyed" by a fisherman who nevertheless shouts "Be careful" when observing a rock, forcing the listener to decide whether to believe the observer or not. He reasons as follows: "My partner has no reason to persuade me that there is an obstacle in my way unless indeed there is an obstacle because:

(1) If I believe him, the best course for me is to prevent a crash, an outcome that is beneficial for him as well.
(2) He gains no benefit from persuading me that there is a risk when there is none."

Farrell's argument is based on the underlying assumption that members of the community understand that the call "Be careful" is meant to indicate the existence of risk. The goal of this research agenda is to explain the existence of a term with this meaning.

2.2 The basic evolutionary argument

The explanation discussed here is provided by evolutionary game theory. The core idea rests on Maynard Smith's (1982) notion of an *Evolutionary Stable Strategy* (ESS), according to which evolutionary forces select the more successful modes of behavior. Language is taken to be a behavioral phenomenon and if it does not serve the needs of the population, evolutionary forces will act to improve its functioning.

If there were no term in the language which allowed one member of the community to alert another, a mutation would occur whereby a small group of mutants would use a certain costless "signal" called "Be careful" to indicate hazard and to correctly respond to such a signal. The mutation can be thought of as originating in nature or a collective decision by a small group of community members. Then, if a mutant is matched with another in a boat, the information regarding the hazard is transferred and utilized. Whether a mutant does better than a nonmutant when he is matched with a nonmutant depends on the way that the cry "Be careful" is interpreted by a nonmutant. If, for example, nonmutants respond to this unfamiliar cry in panic, mutants will, on average, do worse than nonmutated members of the population. In that case, the success of a mutant paired with another mutant would be outweighed by the significant loss incurred when he interacts with a nonmutant (this is the more common case when the mutated population is small). Thus, the explanation of the emergence of a proper meaning for the cry "Be careful" depends on providing a persuasive explanation as to why unfamiliar "words" do not cause "nonmutants" to respond harmfully.

Evolutionary game theory has attempted to resolve this issue. A simple and, in my opinion, very reasonable line of reasoning is presented here from Warneryd (1993) and Banerjee and Weibull (1993). These authors assume that the respective evolutionary forces not only depend on the standard payoffs but also on "strategy complexity costs." If, in equilibrium, no one shouts the phrase "Be careful," then no one would waste his "mental resources" on being prepared to respond to such a cry. It is assumed that evolutionary forces act to minimize the waste of mental resources and eliminate the panic response to the message. Consequently, the term "Be careful" will remain available for mutants to use without suffering a loss from nonmutants' harmful responses.

2.3 The basic situation

Let us now formalize the evolutionary argument outlined in the previous section. We assume a "large" homogeneous population whose members are occasionally paired and examine a situation in which one of them is selected randomly to perceive information about a risk which "must" be transmitted to his partner.

Let W be a nonempty set. We will refer to W as the set of words, although nothing in the model requires the elements of W to be verbal statements. The interaction between the members of the population is modeled as a symmetric game. A strategy in this game has the following format:

(1) When I perceive risk I will use the word $w^* \in W$ or remain silent ("shhh," where "shhh" is not a word in W).
(2) When I hear another person saying $w \in W$, I will take one of three possible actions:
   - "I will take action to avoid the hazard" (A), the correct behavior in the case of a hazard.
   - "I will panic" (P), the worst possible action under any circumstances.
   - I will take the default course of action, D, the best course of action in the case of no hazard and the second best in the case of a hazard.

Thus, a "strategy" for a player in this game has two components: a choice of either "shhh" or $w^* \in W$ and a function $r: W \rightarrow \{D, A, P\}$. Note that no decision is made when there is no hazard.

It is assumed that there is no conflict of interest between any two fishermen so that we are dealing with a coordination game. Given the existence of a hazard, the payoffs depend only on whether or not the listener takes an action to avoid the hazard (payoff 1), remains with the default action (payoff 0), or panics (payoff $-M$, where M is a "large" positive number).

Denote by $u(s,s')$ the expected payoff of a player who uses the strategy s while his partner uses the strategy s', assuming that they have equal chances of being the observer. This definition is extended to $u(s, \pi)$ where π is a probability measure on the set of strategies. A Nash equilibrium in this situation is a strategy s^* satisfying that for no s, $u(s,s^*)>u(s^*,s^*)$. There are two types of equilibria here: In a "separating" equilibrium, a player as an observer of a hazard sends a message w^* and on hearing w^* his response is A. In the pooling equilibria, no message is sent and the listener "plans" to respond to a message by ignoring it or panicking. The pooling equilibria are of course "bad" since valuable information is not being utilized.

This brief presentation raises the question whether there are any evolutionary forces which can bring about the emergence of a "good" equilibrium in which information is transmitted and used.

2.4 Evolutionary forces

The idea that evolutionary forces can explain the emergence of "words" in the natural language appeared in the economic literature as early as Marschak (1965). However, Marschak suggested only a direction for research. It is the literature initiated by Maynard Smith and Price (1973), Maynard Smith (1982), and especially the more contemporary literature of evolutionary game theory, which formulated a coherent argument for the emergence of words. The basic definition is:

Definition 1: A strategy s^* is an Evolutionary Stable Strategy (ESS) if for any strategy s either $u(s,s^*)<u(s^*,s^*)$ or, $u(s,s^*)=u(s^*,s^*)$ and $u(s,s)<u(s^*,s)$.

The rationale for this concept is by now familiar to users of game theory. A mode of behavior s^* will not remain a stable unique mode of behavior in a community if, following a mutation of an arbitrarily small fraction of the population to strategy s, the mutants will not, on average, do worse than the members of the community who have not mutated. The above conditions guarantee that whatever the extent of the mutation, the mutants will do worse than the nonmutants (for every $\varepsilon > 0$ small enough,

$$u(s,\varepsilon s + (1 - \varepsilon)s^*) < u(s^*,\varepsilon s + (1 - \varepsilon)s^*)).$$

This standard form of ESS does not eliminate the pooling equilibria. In fact, if W includes more than one element then no ESS exists. If s^* is an ESS, then any mutation which changes the response to words which are not used will be a "resistible mutation." Thus, for ESS to be a useful concept, we must modify its definition. At least two possible directions have been proposed in the literature.

The first method is to utilize a multi-valued solution concept. The following is one such alternative definition (for others of this type see, for example, Weibull, 1995):

Definition 2: A *set* of strategies S^* is an *evolutionary stable set* if there is a probability measure π on S^* with full support, such that:

(1) $u(s,s') \equiv u^*$ for every $s,s' \in S^*$
(2) For any $s \notin S^*$, $u(s, \pi) < u^*$ or $u(s, \pi) = u^*$ and $u(s,s) < u(s^*,s)$ for any $s^* \in S^*$.

Thus, there is one payoff u^* such that, in an evolutionary stable set, all members of the community receive the same expected payoff u^*. The fact that all of them have the same payoff maintains the stability of the set. The criteria for the success of a mutation are derived from the same rationale as that of the ESS. Mutants will do worse than all strategies which exist in the stable population of strategies. It is easy to verify that the set of all strategies which instructs the player as an observer to remain silent is an evolutionary stable set. Essentially, the placing of a high probability on the strategy which responds to any word by panicking will make any mutant worse off than with any of the S^* strategies.

Note a difficulty with the above definition of evolutionary stable set. Following the appearance of a mutation, not all strategies in S^* will do equally well and thus the distribution of strategies in S^* may be unstable. My impression is that other modifications of the ESS, which appear in the literature, suffer from similar difficulties.

The second approach to modifying the ESS is to weaken it by demanding that a successful mutation must yield strictly higher payoff. This "modified" version (as Binmore and Samuelson, 1992, call it) is appropriate when the mutation is largely intentional and members are interested not in maximizing payoffs but in having the highest payoff in the group:

Definition 3: A strategy s^* is a modified evolutionary stable strategy if for any strategy s either $u(s,s^*)<u(s^*,s^*)$ or $u(s,s^*)=u(s^*,s^*)$ and $u(s,s)\leq u(s^*,s)$.

The strategy dictating the use of the word w^* to warn of a hazard and the response to any message with the action A is a modified evolutionary stable strategy. However, so is the strategy to maintain silence (choose *shhh*) and to respond to any message with panic. Thus, the modified ESS does not provide an explanation of the emergence of a "warning."

2.5 Complexity and evolutionary forces

This final section of the presentation introduces a new element into the model: the complexity of the strategies. It is assumed that evolutionary forces not only affect the basic payoffs but also the complexity of strategies. In particular, evolutionary forces eliminate strategies which demand resources to respond to never-used messages. The costs of complexity put pressure on strategies not to include a "threat" to respond to messages with the harmful action P. This clears the way for mutations that transfer and utilize observed information.

At this point we should formally define what we mean

by the "simplicity of a strategy." A strategy is evaluated by the number of different actions, besides "*shhh*" and "*D*," which it plans to use under certain circumstances even though they will probably not be realized. Denote the complexity of a strategy *s* by comp(*s*). The simplest strategy, with complexity 0, dictates that the player chooses "*shhh*" as an observer and takes the action *D* as a listener whatever the message he receives. It is the simplest strategy in the sense that its execution does not require any resources to monitor the situation. The most complex strategy is of complexity 3. Such a strategy specifies a message to be sent in case of a hazard, a nonempty set of messages to which the response is the action *A* and a nonempty set of messages to which the response is *P*.

As mentioned previously, the basic assumption of this section is that evolutionary forces work in favor of strategies that yield either higher payoffs or lower complexity. It is assumed that the costs of complexity are lexicographically secondary to the payoffs of the original game. Thus, there are two possible modifications of the ESS notion, both of which support the equilibria in which information is transmitted and used.

Definition 4 is analogous to definition 2 with the addition of the complexity considerations:

Definition 4: A *set* of strategies S^* is an *evolutionary stable set* (with complexity considerations) if there is a probability measure π, with the whole set S^* as support, such that:

(1) $u(s,s') \equiv u^*$ for every $s,s' \in S^*$ and all strategies in S^* are equally complex.
(2) For any $s \notin S^*$ $u(s,\pi) < u^*$ or $u(s,\pi) = u^*$ and for all $s^* \in S^*$ either $u(s,s) < u(s^*,s)$ or $[u(s,s) = u(s^*,s)$ and comp$(s^*) < $ comp$(s)]$.

Thus, for a mutated strategy *s* to be successful, either (1) it does better on average than any of the strategies in S^* or, (2) it is equally successful on average as any strategy in S^* but does better than any strategy in S^* when the following two criteria are applied lexicographically: (a) payoff while playing against the mutation s and (b) the complexity of the strategy.

It is clear that there is only one evolutionary stable set (with complexity considerations) which includes any strategy which instructs the user to utter some word in W^* when a hazard is seen and to respond with A to any word in W^*. Thus, this concept "predicts" that information will be transmitted and used, though it has the unattractive feature that any utterance in W^* serves as a meaningful signal.

The final definition is analogous to definition 3:

Definition 5: A strategy s^* is a *modified evolutionary stable strategy with complexity considerations* if for any strategy s either

$u(s,s^*) < u(s^*,s^*)$, or

$[u(s,s^*) = u(s^*,s^*)$ and $u(s,s) < u(s^*,s)]$ or

$[u(s,s^*) = u(s^*,s^*), u(s,s) = u(s^*,s)$ and $comp(s^*) \geq comp(s)]$.

Thus a *modified evolutionary stable strategy (with complexity considerations)* is a modified ESS by which mutant survival is compared using three criteria which are applied lexicographically: the payoff while playing against s^*, the payoff while playing against s, and the complexity of the strategy.

If s^* is such a modified ESS (with complexity considerations) and if s^* assigns "*shhh*" to the observer, then the complexity considerations dictate that the response of s^* to any message is D. It follows that a mutation s which uses the word w to point out a risk and responds to w with A, will be a successful mutation.

If a modified ESS (with complexity considerations), s^*, dictates that w^* be uttered when a hazard is observed then it must be that s^* responds to w^* with A. According to the complexity considerations, there are no circumstances under which any word will trigger the action P and thus s^* must dictate that an observer utters w^* when a hazard is observed and a listener responds with A to any message in some set W^* which contains w^*. It is easy to verify that such a strategy is indeed a modified evolutionary stable strategy (with complexity considerations).

2.6 A final comment

This chapter was meant to provide a short overview of the basic evolutionary game-theoretic explanation for the emergence of words (terms) in natural language. It is debatable whether the model is persuasive. I do not have a firm opinion on this issue. However, I would like to raise three points of criticism:

First, the connection between biological powers and what we model as evolutionary forces is far from being clear. The definitions are not motivated by biological evidence or by intuition.

Second, the scenario analyzed here is not sufficiently general. The emergence of the term "Be careful" is analyzed independently of the other elements of the language. In our analysis we classify the situation of "two members going on a boat trip" as a well defined category. We assume the existence of a well defined meaning for the notion "there is a hazard." We assume that the listener has exactly three possible responses. The only element to be determined in the model is whether the observer shouts a warning when a hazard is observed and whether the shout is interpreted correctly. A persuasive explanation of the emergence of a linguistic concept requires a much more general setting. One may claim that this criticism is applicable to any game-theoretic model. Indeed models of game theory are usually as simplistic and specialized as the one presented here. However, I believe that models of game theory are meant to describe and analyze human reasoning – and do not directly deal with reality. Since people focus on a small number of factors in decision making, game theory is able to use simplified models to explain behavior. In contrast, evolutionary theories are intended to explain real-world phenomena and thus need to be constructed in a more general setting.

Finally, the explanation of such phenomena using evolutionary forces requires an explanation not only for the existence of human language but also for the lack thereof among animals. It is interesting to note that traces of this criticism already appeared in the writings of Adam Smith (see Levy,

1992), for an extensive discussion of this issue. In chapter II of book I, Smith discusses the fact that while human beings conduct trade, "Nobody ever saw a dog make a fair and deliberate exchange of one bone for another with another dog." Adam Smith points out that trade requires language. He attributes the existence of flexible language among human beings to their "hardware", which accounts for the difference between humans and animals. While Adam Smith discusses the link between the evolution of trade and the existence of language, the approach discussed in this chapter has attempted to explain the evolution of language. So what then explains the difference between humans and animals with regard to the development of language? Haven't we simply replaced one puzzle with another?

REFERENCES

Banerjee, A. and J.W. Weibull (1993) "Evolutionary Selection with Discriminating Players," Stockholm: The Industrial Institute for Economic and Social Research

Binmore, K., M. Piccione and L. Samuelson (1998) "Evolutionary Stability in Alternating-offers Bargaining Games," *Journal of Economic Theory*, 80, 257–91

Binmore, K. and L. Samuelson (1992) "Evolutionary Stability in Repeated Games Played by Finite Automata," *Journal of Economic Theory*, 57, 278–305

Blume, A., Y.-G., Kim and J. Sobel (1993), "Evolutionary Stability in Games of Communication," *Games and Economic Behavior*, 5, 547–75

Farrell, J. (1993) "Meaning and Credibility in Cheap-talk Games," *Games and Economic Behavior*, 5, 514–31

Levy, D. (1992) *The Economic Ideas of Ordinary People*, London: Routledge

Marschak, J. (1965) "The Economics of Language," *Behavioral Science*, 10, 135–40

Maynard Smith, J. (1982) *Evolution and the Theory of Games*, Cambridge: Cambridge University Press

Maynard Smith, J. and G.R. Price (1973) "The Logic of Animal Conflict," *Science*, 246, 15–18

Rubinstein, A. (1986) "Finite Automata Play the Repeated Prisoner's Dilemma," *Journal of Economic Theory*, 39, 83–96

Warneryd, K. (1993) "Cheap Talk, Coordination and Economic Stability," *Games and Economic Behavior*, 5, 532–46

(1998) "Communication, Complexity and Evolutionary Stability," *International Journal of Game Theory*, 27, 599–609

Weibull, J. (1995) *Evolutionary Game Theory*, Cambridge, MA: MIT Press

CHAPTER 3

STRATEGIC CONSIDERATIONS IN PRAGMATICS

3.1 Grice's principles and game theory

The study of languages (see Levinson, 1983) is traditionally divided into three domains: *syntax, semantics,* and *pragmatics.* Syntax is the study of language as a collection of symbols detached from their interpretation. Semantics studies the rules by which an interpretation is assigned to a sentence independently of the context in which the sentence is uttered. This chapter deals with the domain of pragmatics.

Pragmatics examines the influence of *context* on the interpretation of an utterance. An "utterance" is viewed as a signal which conveys information within a context. The context comprises the speaker, the hearer, the place, the time, and so forth. How the hearer views the intentions of the speaker and how the speaker views the presuppositions of the hearer are relevant to the understanding of an utterance. Thus, in game-theoretic terms, the way in which an utterance is commonly understood may be thought of as an equilibrium outcome of a game between speakers of a language.

To illustrate the type of phenomena explained by pragmatics, consider the following natural language conversations:

Example 1: A telephone conversation between *A* at home and his friend *B*, calling from a telephone booth:

Sections 3.2–3.4 are based on Glazer and Rubinstein (1997). For basic readings on pragmatics see Levinson (1983). Grice (1989) is a collection of Grice's papers on the subject.

A: "*B*, I am about to go for a walk. What is the weather like outside?"

B: "It is not raining heavily now."

Normally, *A* concludes from *B*'s statement that it is raining but not heavily. This conclusion does not follow from the semantic interpretation of the sentence "it's not raining heavily," which allows for the possibility that it is "raining but not raining heavily" as well as "it's not raining at all." Furthermore, there are circumstances in which the utterance "it's not raining heavily" will indeed be interpreted as not excluding the possibility that it is not raining at all. For example, imagine that *B* is in a hut and has told *A* that it is dark outside and that he is buried under his blanket. In such a case, *B* would only be aware of rain that is "pouring down" and hitting the roof loudly. In this case *A* could infer only that it is not raining heavily from *B*'s statement since *B* would not know whether it was raining lightly or not raining at all.

Example 2:

A: "Where does *C* live? I need to reach him urgently."

B: "*C* is somewhere in New York."

Unless *A* suspects that *B* is trying to prevent him from reaching *C*, *B* will normally conclude that *A* really does not know *C*'s exact address, although he did not say so explicitly.

Example 3:

A, a married man, to *B*: "I went out with a woman last night."

A's statement is normally understood to mean that he dated someone who is not his wife although he did not explicitly say so.

Example 4:

A: "What do you see?"

B: "It is not a rose."

Ordinarily, *A* will understand that *B* is looking at a flower which is not a rose, although *B* has not said that he is looking at a flower.

In these examples, the fact that a statement appeared within a conversation between two people gives the sentence a different meaning than if it were interpreted in isolation. A theory which attempts to describe the rules for interpreting conversational utterances was suggested by the philosopher Paul Grice in the 1960s (see especially his article "Logic and Conversation" in Grice, 1989) as follows:

> while it is no doubt true that the formal devices are especially amenable to systematic treatment by the logician, it remains the case that there are very many inferences and arguments, expressed in natural language and not in terms of these devices, which are nevertheless recognizably valid. So there must be a place for unsimplified ... logic of the natural counterparts of these devices. (pp. 23–4)

According to Grice (1989), speakers of a language are guided by "the cooperative principle":

> Make your conversational contribution such as is required, at the stage at which it occurs, by the accepted purposes or direction of the talk exchange in which you are engaged. (p. 26)

Grice suggests four "maxims" which "in general yield results in accordance with the cooperative principle":

Quantity: "Make your contribution as informative as is required ... Don't make your contribution more informative than is required."

Quality: "Try to make your contribution one that is true ... 1. Do not say what you believe to be false. 2. Do not say that for which you lack adequate evidence."

Relation: "Be relevant."

Manner: "Be perspicuous (avoid obscurity, avoid ambiguity, be brief and be orderly)."

A central concept in Grice's theory is that of *conversational implicature:*

A man who, by saying that *p* has implicated *q*, may be said to have *conversationally implicated* that *q*, provided that (1) he is presumed to be observing the conversational maxims, or at least the Cooperative Principle; (2) the supposition that he is aware that, or thinks that, *q* is required in order to make his saying or making as if to say *p*, consistent with this presumption; and (3) the speaker thinks (and would expect the hearer to think that the speaker thinks) that it is within the competence of the hearer to work out, or grasp intuitively, that the supposition mentioned in (2) is required. (1989, pp. 30–1)

The inference mechanism which is used in conversational implicature is described by Grice as follows:

He had said that *p*; there is no reason to suppose that he is not observing the maxims, or at least the Cooperative Principle; he could not be doing this unless he thought that *q*; he knows (and knows that I know that he knows) that I can see that the supposition that he thinks that *q* is required; he has done nothing to stop me from thinking that *q*; he intends me to think, or is at least willing to allow me to think, that *q*; and so he has implicated that *q*. (1989, p. 31)

Following are some examples of how Grice's theory is applied:

Example 5: *A* is a driver who stops in a remote village and sees *B* who looks clearly a local.

A: "I am out of petrol."
B: "There is a garage around the corner."

A's line of thought might be as follows:

- There is no reason for me not to believe that *B* is co-operative.
- *B*'s utterance must not interfere with the maxim "be relevant."
- It must be that *B* means that petrol is sold at that garage.

Let us return to example 2: Unless *B* wants to prevent *A* from reaching *C*, the maxim of quality seems to be violated, unless *B* does not know *C*'s address. In example 3, the hearer excludes the possibility that the speaker does not know the name of the woman he was dating last night. The fact that A does not know who the woman is must indicate that he does not want to give the name to B, which normally implies that she is not his wife.

Note that Grice's inference does not apply to circumstances in which it is clear that the relationship between the speaker and the hearer is not cooperative.

Example 6: A and B play a zero-sum game: B hides a coin in one of his two hands. A will get the coin if he correctly guesses the hand holding the coin.

A: "Hmmm ..."
B: "The coin is in my right hand."

Here, the cooperative principle clearly does not apply. A would normally conclude that B's statement is an attempt to confuse him.

Grice's explanation is not always persuasive. Let us return to example 1:

A: "B, I am about to go for a walk. What is the weather like outside?"
B: "It is not raining heavily now."

For A to conclude that B means that it is raining but not heavily, A has to go through the following steps:

- B knows more than he has told me. He knows the state of the world exactly, which may be "not raining" (NR), "light rain" (L), or "heavy rain" (H).
- He wants to be cooperative and is giving me true information.
- He wants to be relevant so he does not give me the exact information.
- If it was not raining at all, he would tell me so.
- Since he says that it is not H, and if it were NR he would tell me, I conclude that the state of the world is L.

The weak link in this chain is A's assumption that if B were to observe that it was not raining, he would convey this information to A. If B wants to be relevant and "it is not raining heavily" is the relevant information, why would he not say so if it were not raining at all?

Students of game theory will find this discussion to be in the spirit of game-theoretic reasoning. Note, in particular, the considerations which Grice imposes on the listener: "he (the speaker) could not be doing this unless he thought that q; he knows (and knows that I know that he knows) that I can see that the supposition that he thinks that q, is required ..." In other words, Grice's theory is essentially a description of one agent thinking about how another is thinking. This is precisely the definition of strategic reasoning and is the essence of game theory (for an early paper which gives an account of Gricean communication using the framework of game theory see Parikh, 1991).

As Grice noted, the theory may also be applicable to other forms of interactions between informed and uninformed agents, in which the informed agent takes an action (not necessarily an utterance) with the intention of influencing the hearer's subsequent actions. The similarity to game theory is clear.

Furthermore, if game theory is to shed light on real-life phenomena, linguistic phenomena are the most promising candidates. Game-theoretical solution concepts are most suited to stable real-life situations which are "played" often by large populations of players. Thus, game theoretic tools may function most effectively when used to explain linguistic phenomena.

3.2 Debates

To demonstrate the possible applications of game-theoretic methods in explaining Gricean phenomena, the remainder of this chapter will focus on a phenomenon related to debates. In this chapter, a "debate" will refer to a situation in which two parties who disagree regarding some issue raise arguments in an attempt to persuade a third party of their positions. This is, of course, not the only possible form of a debate. In some cases, the intention of the debaters is to argue just for the sake of arguing. In other cases, their aim is to influence one another rather than a third party. Debates can be thought of as a special form of conversation in which the parties have different interests. Debates differ from other mechanisms of "conflict resolution," such as bargaining and war, since power is not used (or is used less) in debates to influence the outcome. Although debates are common in real life, they have rarely been investigated within the economic and game theoretic literature (notable exceptions are Shin, 1994; Lipman and Seppi, 1995 and Spector, 2000).

The aim of this chapter is to provide an explanation of the following phenomenon, often observed in debates: Two arguments may have a different degree of persuasiveness as counter-arguments even though they would be viewed as equally persuasive when stated in isolation. At the outset, we should point out that the Gricean theory does not apply to debates since Grice's logic of conversation is based on the principle of cooperation, which does not apply to the conflict of interests characteristic of a debate.

To initiate the discussion, the results of a survey conducted among several groups of students at Tel-Aviv University will be presented. The following question was presented to one group of students:

Question 1: You are participating in a public debate on the level of education in the world's capitals. You are trying to convince the audience that in most capital cities, the level of education has recently increased. Your opponent is challenging you with indisputable evidence showing that the level of education in *Bangkok* has deteriorated. Now it is your turn to respond. You have similar indisputable evidence to show that the level of education in *Mexico City, Manila, Cairo*, and *Brussels* has increased. However, owing to time constraints, you can present evidence for only one of the four cities. Which city would you choose to make the strongest counter-argument to the *Bangkok* results?

Another group of subjects was presented with question 1 with the modification that Bangkok was replaced by Amsterdam. To verify that the subjects regarded the four potential counter-arguments as equally effective, a third group of students was asked to answer question 2, in which the subjects are asked to select an opening argument rather than a counter-argument:

Question 2: You are participating in a public debate on the level of education in the world's capitals. You are trying to convince the audience that in most capital cities, the level of education has recently increased. You have indisputable evidence showing that the level of education in *Mexico City, Manila, Cairo* and *Brussels* has improved. Owing to time constraints, you can present evidence for only one of these cities. Which city would you choose in order to make your argument as convincing as possible?

The following table presents the responses to questions 1 and 2. Manila is the favorite counter-argument to Bangkok while Brussels is the favorite counter-argument to Amsterdam. In contrast, the subjects split quite evenly between the four cities in question 2.

	Question 1 Bangkok	Question 1 Amsterdam	Question 2
n	38	62	24
Mexico City (%)	19	15	21
Manila (%)	50	3	21
Cairo (%)	11	5	25
Brussels (%)	21	78	33

A different group of students responded to question 3:

Question 3: Two TV channels have fixed weekly program schedules. You and a friend are debating which is the better channel in the presence of a third party. Your opponent argues in favor of channel *A* while you argue in favor of channel *B*. Both of you has access to the same five reports prepared by a highly respected expert, each of which refers to a different day of the week and recommends one channel over the other for that day. Your opponent begins the debate by quoting *Tuesday*'s report, which recommends channel *A*'s programs. The third party interrupts him and asks you to reply. Both *Wednesday*'s and *Thursday*'s reports are in your favor – i.e. they recommend Channel *B*. However, you have time to present only one of these reports as a counter-argument, following which the third party will make up his mind. Would you choose the report for Wednesday or for Thursday as a better counter-argument to Tuesday?

The results were quite clear: About 70 percent of the 58 subjects felt that Wednesday would be a better counter-argument than Thursday. Note that Thursday is the last weekday in Israel and hence, one would expect it to be more significant than Wednesday.

The survey results are puzzling. If two arguments contain information of the same quality, why is one considered to be a stronger counter-argument than the other? The fact that Manila is closer to Bangkok than it is to Mexico City seems irrelevant to the substance of the debate, yet it appears to dramatically affect the choice of the better counter-argument. Similarly, although Wednesday is less significant than Thursday with regard to TV programming, it is assessed as a better counter-argument against Tuesday.

We believe that this phenomenon is connected to pragmatics. A debater's response to "Bangkok" with any counter-argument other than Manila is interpreted as an admission that Manila is also an argument in favor of the opponent's position. In the context of question 3, if one responds to Tuesday with the counter-argument Thursday, rather than Wednesday, it is considered an admission that Wednesday's report is not in his favor. The fact that the sentence "On Thursday Channel *B* is superior" is uttered in a debate as a counter-argument to "On Tuesday Channel *A* is superior" gives the sentence a meaning different than if the statement were made in isolation.

In what follows, I have no pretensions to fully explain the rules by which people compare arguments. The sole target is to point out that the logic of debate should not necessarily include an axiom stating that if an argument x is superior to argument y (in the sense of being a successful counter-argument against it) then argument y should not be superior to argument x.

3.3 The model

We view a debate as a mechanism designed to extract information from debaters. We assume that the aim of the designer is to increase the probability that the right conclusion will be drawn by the listener subject to the constraints imposed on the debaters and the listener in terms of time and cognitive abilities. It will be shown that fulfilling this aim may lead to debating rules which do not treat arguments and counter-arguments symmetrically.

The setup is very simple: An uninformed *listener* has to choose between two *outcomes*, O_1 and O_2. The "correct" outcome, from his point of view, is determined by five *aspects*, numbered 1, ..., 5. An aspect i may be realized as either 1 or 2. When an aspect i receives the value j it implies that aspect i possibly supports the outcome O_j $(j=1,2)$. A *state* $\omega = (\omega_j)_{j=1,...,5}$ is a five-tuple of 1s and 2s which describes the realizations of the five aspects. The listener assigns equal weights to all five aspects, and the *correct* outcome in state ω, $C(\omega)$, is the one which is supported by the majority of the arguments.

While the listener is ignorant of the state, the two debaters, named *debater 1* and *debater 2*, have full information about the state. There exists a conflict of interests between the two debaters and the listener: Each debater i wishes that outcome O_i be chosen, whatever the state, whereas the listener wants the correct outcome to be chosen.

A debate is defined to be a mechanism in which each debater reveals information in order to persuade the listener to choose his preferred outcome. A debate consists of two elements: The *procedural rule* which specifies the order and types of arguments which are permitted and the *persuasion rule* which specifies the relationship between the arguments presented and the listener's conclusion.

The "language" that the debaters are permitted to use must be specified. It is assumed that debaters cannot make any moves other than raising arguments of the type "argument j supports me." Thus, a debater cannot raise arguments that support the outcome preferred by the other debater. We will further assume that the debaters must prove their claims – i.e. debater i cannot claim that the value of aspect j is i unless it is indeed so.

The optimal debate should have an effective constraint on its length. Of course, if it is possible for three arguments to be raised during the debate (that is, there is enough time for raising three arguments and the listener has the cognitive ability to digest them all), then the listener can obtain the correct outcome with certainty. He can accomplish this by requesting that one of the debaters present three arguments – the debater will win the debate if and only if he fulfills this task. Thus, for the length constraint to be effective, it is assumed that the number of arguments which can be raised during a debate is constrained to two.

Formally, we define a *debate* to be an extensive game form in which:

(1) The set of feasible moves for each debater is a subset of $\{1, \dots, 5\}$.

(2) There can be at most two moves: Either only one of the debaters is allowed to make at most two arguments (the *one-speaker debate*); or the two debaters move simultaneously, each one making at most one argument (the *simultaneous debate*); or the debate is a two-stage process in which one debater is allowed to make at most one argument in each stage (the *sequential debate*).

(3) One of the outcomes, O_1 or O_2, is attached to each terminal history.

When debater j presents an argument to answer a previous argument of debater i, debater j's argument is referred to as a *counter-argument*.

A debate Γ and a state ω determine a game $\Gamma(\omega)$ which will be played by the two debaters when the state is ω. The two-player game form $\Gamma(\omega)$ is obtained from Γ by deleting, for each debater i, all aspects which do not support his position in ω. If player i has to move following a history h and if at ω none of the arguments he is allowed to make at

h support his position, then the history h in the game $\Gamma(\omega)$ will become a terminal history and the outcome O_j (debater i loses) is attached to h. As to the preferences in $\Gamma(\omega)$, debater i strictly prefers outcome O_i to O_j. The game $\Gamma(\omega)$ is a two-person game; the listener is not considered a player. The game is a zero-sum game and has a value, $v(\Gamma,\omega)$, which is a *lottery* over the set of outcomes satisfying that each debater i has a strategy such that whatever the other debater's strategy is, the distribution of outcomes is at least as good for i as this lottery. Let $m(\Gamma,\omega)$ be the probability that $v(\Gamma,\omega)$ assigns to the incorrect outcome. When $m(\Gamma,\omega)=1$, we say that debate Γ induces a *mistake* in state ω. In debates which are not simultaneous, $m(\Gamma,\omega)$ is either 0 or 1. In simultaneous debates, $m(\Gamma,\omega)$ may be a nondegenerate lottery $(1 > m(\Gamma,\omega) > 0)$. For example, consider the simultaneous debate Γ with the persuasion rule by which debater 1 wins if and only if he argues for some i and debater 2 does not argue for either $i+1(\mathrm{mod}5)$ or $i-1(\mathrm{mod}5)$. For the state $\omega = (1, 1, 2, 2, 2)$, the value of the game $\Gamma(\omega)$ is O_i with probability ½ for both i; thus $m(\Gamma,\omega)=1/2$.

All mistakes are weighted equally and the *optimal debate* is taken to be the one which minimizes $m(\Gamma) = \Sigma_\omega m(\Gamma,\omega)$. The following examples demonstrate the calculation of $m(\Gamma)$:

(1) Let Γ be the debate in which only debater 1 is allowed to speak and the listener is persuaded by any two arguments that support debater 1's position. Then, a mistake occurs in any state ω where the number of aspects which support debater 1 is exactly 2 – that is, $m(\Gamma)=10$.

(2) Let Γ be a debate in which only debater 1 is asked to present two arguments and he wins the debate if he raises two arguments in his favor from the set $\{1,2,3\}$ or the set $\{4,5\}$. In this case $m(\Gamma)=4$: the four mistakes in favor of debater 1 occur in states where only aspects 4 and 5 or two aspects out of the set $\{1,2,3\}$ support debater 1. In fact, this debate has the least number of mistakes in the set of one-speaker debates.

(3) Let Γ be a simultaneous debate in which each debater can present at most one argument. Debater 2 wins the debate if debater 1 uses aspect x and he responds with a counter-argument regarding aspect $x + 1 \pmod 5$ or aspect $x - 1 \pmod 5$. In this case, debater 1 rightly wins in any state ω where there are three successive aspects (ordered on a circle) in his favor and he loses in any state where only zero, one or two nonconsecutive arguments are in his favor. There are ten other states: five of them are the "shift permutation" of $(1, 1, 2, 1, 2)$ and the other five are the "shift permutation" of $(1, 1, 2, 2, 2)$. In each of these ten states, the value of the induced game is the lottery that selects the two outcomes with equal probabilities. Thus, $m(\Gamma) = 10(\frac{1}{2}) = 5$. In fact, one can see that the minimal number of mistakes in the family of simultaneous debates is 5.

(4) The following is a sequential debate which induces three mistakes. First, debater 1 and then debater 2 are asked to present an argument. Debater 2 wins if and only if he counter-argues argument i with an argument regarding an aspect which is listed in the second column in row i. This debate induces three mistakes, two in favor of debater 1 (in states (1,1,2,2,2) and (2,2,1,1,2)) and one in favor of debater 2 (in state (1,2,1,2,1)).

If debater 1 argues for ...	Debater 2 wins if and only if he counter argues with ...
1	{2}
2	{4,5}
3	{4}
4	{1,5}
5	{2,3}

Actually, one can show that this is an optimal debate (see Glazer and Rubinstein, 1997). To conclude:

Claim: Any optimal debate procedure is sequential. The minimal $m(\Gamma)$ over all debates is three.

Let us continue discussing the above optimal debate. What does the listener understand from an exchange of arguments in which debater 1 argues regarding aspect i and debater 2 responds with aspect j? Is it merely that aspect i supports debater 1 and aspect j supports debater 2? We can analyze the listener's thought process by considering the sequential equilibria of the three-player game constructed from the debate game by adding the listener as a third player who has to choose the outcome of the debate at any terminal history. The above optimal persuasion rule is supported by a sequential equilibrium in which debater 1's strategy is to raise the first argument i for which debater 2 does not have a proper counter-argument. If debater 2 has a proper counter-argument for each of his feasible arguments, debater 1 chooses the first argument which supports him. Debater 2's strategy is to respond with a successful counter-argument whenever possible. The listener chooses the outcome according to the above persuasion rule. Note, for example, that if debater 1 raises argument 3 and debater 2 raises argument 4, it is optimal for the listener to rule in favor of debater 2 since he concludes that in addition to aspect 4, aspects 1 and 2 are in favor of debater 2.

Note, however, that a three-player sequential debate game has other sequential equilibria. One of those is particularly intuitive: In any state ω, debater 1 raises the first argument i which is in his favor and debater 2 responds with the argument j, which is the smallest $j > i$ in his favor. The listener's strategy is guided by the following logic: In equilibrium, debater 1 is supposed to raise the first argument in his favor. If he raises argument i, the listener then believes that arguments $1, 2, \ldots, i-1$ are in favor of debater 2. In equilibrium, debater 2 is supposed to raise the first argument in his favor following argument i. Hence, if debater 2 raises argument j, the listener believes that arguments $i+1, \ldots, j-1$ are in favor of debater 1. The listener chooses the outcome supported by the debater who he believes has more aspects in his favor; in the case of a tie, he rules in favor of debater 2 (stated differently, debater 2 wins if $(i-1)+1 \geq (j-i)$). This equilibrium induces six mistakes, in the states $(1, 1, 2, 2, 2)$, $(1, 2, 2, 1, 1)$, $(1, 2, 1, 2, 1)$, $(1, 2, 1, 1, 2)$, $(2, 1, 2, 1, 1)$, and $(2, 1, 1, 2, 1)$.

It is interesting that according to the above optimal persuasion rule, aspect 5 is a persuasive counter-argument against aspect 2 and aspect 2 is a persuasive counter-argument against debater 1's argument regarding aspect 5. This is not coincidental. In fact we can draw the following conclusion (see Glazer and Rubinstein, 1997):

Conclusion: Any optimal debate is sequential and has a persuasion rule which does not treat the players symmetrically. In other words, there is a pair of aspects, i and j, such that when presented in sequence, i is a persuasive counter-argument against j and j is a persuasive counter-argument against i.

3.4 Discussion

The discussion in this chapter is not intended to provide an explanation for the precise manner in which statements within a debate are interpreted. It demonstrates a rationale for a property which is often observed in real life: The strength of arguments when used as counter-arguments may be different than when they are used as arguments. The explanation is that persuasion rules are intended to create an environment enabling the best elicitation of information, given the constraints on the length of the debate and the interests of the debaters. Thus, the fact that argument i defeats argument j and, at the same time, argument j defeats argument i, is not necessarily a contradiction. In other words, the "logic of debate" does not include, and does not *have* to include, a rule that if "p defeats q," "q should not defeat p."

Persuasion rules in natural language have to be stated in natural terms. The natural language terms in the "five cities" example are the "relative geographical distance" and "geographical partition" of the group of cities into the set of Far Eastern cities {Bangkok, Manila} and non-Far Eastern cities {Cairo, Brussels, Mexico City}. The optimal three-mistake persuasion rule that was described in the previous section is not definable in these terms and thus it is not reasonable to expect that we would observe it in real life. On the other hand, the following is an example of a persuasion rule for a sequential debate for the "five cities" example which can be described by the "Far East" and "non-Far East" terms: If the first speaker used a Far Eastern city in his argument, then the second must answer with a Far Eastern city, and similarly for a non-Far Eastern city. The sequential debate with this persuasion rule has seven mistakes (one in the state where only Bangkok and Manila support the first speaker and the other six in each state where exactly two of the non-Far Eastern cities and exactly one of the Far Eastern cities support the first speaker). This geographical partition also allows for a four-mistake one-speaker debate in which the speaker is required to present two arguments in order to win, either from the set of Far Eastern cities or from the set of non-Far Eastern cities. Thus, the conclusion that there is a sequential debate which is superior to all other forms of debate is sensitive to the language in which the debate rules can be stated.

REFERENCES

Glazer, J. and A. Rubinstein (1997) "Debates and Decisions: On a Rationale of Argumentation Rules," Tel Aviv University, mimeo

Grice, P. (1989) *Studies in the Way of Words*, Cambridge, MA: Harvard University Press

Levinson, S.C. (1983) *Pragmatics*, Cambridge: Cambridge University Press

Lipman, B.L. and D.J. Seppi (1995) "Robust Inference in Communication Games with Partial Provability," *Journal of Economic Theory*, 66, 370–405

Parikh, P. (1991) "Communication and Strategic Inference," *Linguistics and Philosophy*, 14, 473–514

Spector, D. (2000) "Rational Debate and One-dimensional Conflict," *Quarterly Journal of Economics*, 115

Shin, H.S. (1994) "The Burden of Proof in a Game of Persuasion," *Journal of Economic Theory*, 64, 253–64

PART 2

LANGUAGE OF ECONOMICS

CHAPTER 4

DECISION MAKING AND LANGUAGE

4.1 Introduction

The starting point of this chapter is the view that decision makers deliberating before making a choice often verbalize their considerations. It follows that the "language" a decision maker uses to verbalize his preferences restricts the set of preferences he may hold. Thus, interesting restrictions on the richness of the decision maker's language can yield interesting restrictions on the set of an economic agent's admissible preferences.

Before we proceed to the main investigation, some background is required. An economic agent in a standard economic model possesses a preference relation defined on a set of relevant consequences. The preferences provide the basis for the systematic description of his behavior as well as for welfare analysis. We usually assume that an economic agent is "rational" in the sense that his choice derives from an optimization given his preferences. Given that we adopt the rational man paradigm, the other constraints imposed upon an economic agent's preferences are often weak. For example, in general equilibrium theory, we usually only impose conditions such as monotonicity, continuity, and quasi-convexity. On the other hand, we restrict the discussion in many economic studies to some

The basic ideas in this chapter appeared in Rubinstein (1978), one of my early papers which was never published. A related paper is Rubinstein (1984). Two parts of this chapter appeared as Rubinstein (1998b) and Rubinstein (1999).

The mathematical tools I will be using are from standard Mathematical Logic (Boolos and Jeffrey, 1989, and Crossley *et al.*, 1990, are good introductory books).

family of preference relations which share a simple utility representation. Several questions thus arise: What is the reason that we do not further restrict the set of preferences in the many cases where generality prevents us from obtaining stronger results? Why do we often restrict our attention to a small family of preferences even though the results we obtain depend heavily on this restriction? To phrase it more provocatively: *Why does the utility function* $(log(x_1 + 1))x_2$ *in a two-commodity world lie within the scope of classic textbooks whereas lexicographic preferences do not?*

What are the general considerations that motivate us to include or exclude preferences from the scope of analysis? One consideration is that some preferences may better explain empirical data. Although I am not an empirical economist, I am doubtful that this is a factor in the choice of restrictions imposed on preferences in the economic theory literature. Another consideration is "analytical convenience." This is a legitimate consideration, but one which must be treated with the necessary caution. Another consideration relates to "bounded rationality." It may be argued that some preferences are more plausible than others since they can be derived from plausible procedures of choice. Finding such derivations is, of course, one of the main objectives of bounded rationality models (for a discussion of this point see chapter 2 in Rubinstein, 1998a).

This analysis, however, will consider a different issue: the availability of a description of the preferences in a decision maker's language. The chapter assumes that when a decision maker is involved in an intentional choice, he describes his considerations, to himself or to other agents who operate on his behalf, using his daily language. (For a discussion of this point in the psychological literature see, for example, Shelly and Bryan, 1964.) Thus, "My first priority is to obtain as many guns as possible and only after that do I worry about increasing the quantity of food" is a natural description of a preference relation. "I spend 35 percent of my income on food and 65 percent on guns" is a natural description of a rule of behavior (one consistent with maximizing some Cobb–Douglas utility function). The function $(\log(x_1 + 1))x_2$, on the other hand, is a "textbook utility function" that is expressed by a relatively simple mathematical formula, even though there is no rule of behavior, stated in everyday language, which corresponds to this utility function.

Note that when a decision maker is a collective (recall that decision makers in economics are often families, groups, or organizations) the assumption that preferences must be definable makes even more sense. In this case, a decision rule must be stated in words in order to be communicated among the individuals in the collective, during both the deliberation and the implementation stages.

The assumption that the decision maker uses a limited language to express his preferences can have different interpretations. Most naively, it may be interpreted as a reflection of the limited language that is available to the decision maker. But even if the language available to the decision maker is rich, he may not be able to use the richer language because the relevant information he possesses is limited. And even if the language is richer and even if more information is available, the decision maker may choose, because of complexity considerations, to express his preferences by a formula, which must be stated in a limited language.

The aim of this chapter is to demonstrate that the requirement that preferences be definable can be fruitfully analyzed in formal terms. More specifically, we will analyze two examples which illustrate the connection between a decision maker's language and the set of definable preferences. This analysis may serve as the first step in a much more ambitious research program to study the interaction between economic agents with "language" as a constraint on agents' behavior, institutions, communication, etc. However, dreams aside, the goal of this chapter is quite modest.

4.2 Definable preferences on the basis of binary relations

Preferences are often defined on the basis of primitive binary relations. For example, it is common that a decision maker deliberating over the choice between two cars will state his considerations explicitly and say something like: "I have three criteria in mind – price, safety, and the consumer report ranking. I prefer one car over the other since according to most of these criteria, it is the superior car." When these binary relations are preferences, this aggregation is an operation that we call a "social rule" in economics. The study of this operation is at the core of social choice theory in which basic preferences are interpreted as individualistic preference relations and aggregate preferences are those of the collective as a whole. If one replaces the term "individual i" with "property i," social choice theory is transformed from a theory of social decisions into a theory of the formation of individualistic preferences.

The assumption modeled and discussed in this section is that in comparing two alternatives, the decision maker uses a formula that includes the names of the K binary relations, but not the names of the alternatives. The decision maker is assumed to have a "rule" that determines his preference between any pair of elements without referring to their names. His preference is expressed by a sentence involving only phrases about the way that the K primitive relations relate "one element" to the "other."

In order to proceed with the analysis we must first specify the language formally. We will use the *language of propositional calculus* for this purpose. For every k, denote by $[xP_ky]$ the atomic proposition, which is interpreted as "the element x relates to the element y according to the kth binary relation." Note that $[xP_ky]$ is just a symbol representing a statement such as "one car is more expensive than the other."

Recall that a formula in the language of the calculus of propositions, with "atomic propositions" $v_1, \ldots, v_L$, is a string of symbols constructed inductively by the following rules: Any atomic proposition is a formula and, if φ and ψ are formulae, then $-\varphi$, $\varphi \wedge \psi$, $\varphi \vee \psi$, $\varphi \rightarrow \psi$, and $\varphi \leftrightarrow \psi$ are also formulae. The truth value of such a formula φ – i.e. when v_k receives the truth value t_k ("True" or "False") – is denoted by $\varphi(t_1, \ldots, t_L)$ and is defined inductively using the standard truth tables of the connectives.

Let A be a set of alternatives. We will refer to a function that assigns a binary relation on A to any profile of K-tuples of binary relations on A as an *aggregation rule*. An aggregation rule is attached to every formula with the atomic propositions $[xP_1y], ..., [xP_Ky]$:

Definition: An aggregation rule R is *defined* by φ, a formula in the calculus of propositions with the atomic propositions $[xP_1y], ..., [xP_Ky]$, if, *for any a, $b \in A$ and for every profile of binary relations $(R_1, ..., R_K)$, $aR(R_1, ..., R_K)b$ if and only if $\varphi(aR_1b, ..., aR_Kb) = T$.*
If such a formula φ exists, then the aggregation rule R is *definable*.

In other words, the determination whether $aR(R_1, ..., R_K)b$ is true or false is carried out according to the following process: First, calculate the truth values of the K variables $\{aR_1b, ..., aR_Kb\}$; then calculate the truth value of φ when the truth value of the atomic proposition $[xP_ky]$ is set to the truth value of aR_kb. Note that the same formula φ is used to determine the truth of $aR(R_1, ..., R_K)b$ for all pairs $a, b \in A$ and for all possible profiles of primitive binary relations $(R_1, ..., R_K)$.

Two examples: The conjunction $\bigwedge_{k=1, ..., K}[xP_ky]$ states the Pareto principle – one element relates to the other only if x relates to y by all K criteria. The majority rule can be written as the disjunction of all formulae of the type $\bigwedge_{k \in M}xP_ky$, with $M \subseteq \{1, ..., K\}$ containing more than $K/2$ elements.

Note that the requirement that in this case the aggregation rule R be definable can be replaced by much more standard requirements:

For any two profiles of binary relations $\{R_k'\}_{k=1, ..., K}$ and $\{R_k''\}_{k=1, ..., K}$ and for any two pairs of alternatives (a, b) and (c, d) of elements in A, if for all k $[aR_k'b$ if and only if $cR_k''d]$ then $aR(R_1', ..., R_K')b$ if and only if $cR(R_1'', ..., R_K'')d$.

This condition has three components which are familiar from social choice theory:

(0) The rule which determines whether aRb or not aRb depends only on $\{aR_kb\}_{k=1,\ldots,K}$; no other information is relevant (such as whether bR_ka).

(1) Given any profile of binary relations, the same rule determines the "aggregated" relation between any pair of alternatives. This condition is in the spirit of the *neutrality* condition in social choice theory.

(2) For any pair of alternatives a and b, the same rule determines how a relates to b for all profiles of primitive relations. This condition is in the spirit of the *Independence of Irrelevant Alternatives* in social choice theory.

It is not always this simple to identify a standard condition that is equivalent to a definability condition. In any case, the "definability" approach gives the standard conditions a more natural interpretation.

The binary relations in the domain of an aggregation rule are often naturally restricted to satisfy certain properties which are systematic dependencies of the truth value of aR_kc on the truth values of aR_kb and bR_kc, for any a,b, and c. Such a requirement can be written as a formula, T_k, with the atomic propositions $[xP_ky]$, $[yP_kz]$, and $[xP_kz]$, in the form of a noncontradicting conjunction of some of the eight formulae $\delta_1[xP_ky]\wedge\delta_2[yP_kz]\to\delta_3[xP_kz]$, where each δ_i is either -1 or $+1$ (with the notation $-1\varphi\equiv\neg\varphi$ and $+1\varphi\equiv\varphi$).

We say that a formula T_k is *deterministic* if T_k is a conjunction of exactly four formulae of the type $\delta_1[xP_ky]\wedge\delta_2[yP_kz]\to\delta_3[xP_kz]$, one for each of the possible configurations of δ_1 and δ_2. If T_k is deterministic, the truth value of aR_kc is fully determined by the truth values of aR_kb and bR_kc (for any a,b, and c).

We say that T_k is *degenerate* if there is some δ_1 such that $\delta_1[xP_ky]$ implies $\delta_3[xP_kz]$ (or there is some δ_2 such that $\delta_2[yP_kz]$ implies $\delta_3[xP_kz]$) for some δ_3 – or, more precisely, there is some δ_1 such that both $\delta_1[xP_ky]\wedge[yP_kz]\to\delta_3[xP_kz]$ and $\delta_1[xP_ky]\wedge-[yP_kz]\to\delta_3[xP_kz]$ are conjuncts in T_k, or some δ_2 such that both $[xP_ky]\wedge\delta_2[yP_kz]\to\delta_3[xP_kz]$ and $\delta_1[xP_ky]\wedge-\delta_1[yP_kz]\to\delta_3[xP_kz]$ are conjuncts in T_k.

For example, the formula $([xP_ky]\wedge[yP_kz]\to[xP_kz])\wedge$ $(\neg[xP_ky]\wedge\neg[yP_kz]\to\neg[xP_kz])$ corresponds to the requirement that the kth primitive relation be a linear ordering. The formula $([xP_ky]\wedge[yP_kz]\to[xP_kz])\wedge([xP_ky]\wedge\neg[yP_kz]\to\neg$ $[xP_kz])\wedge\neg[xP_ky]\wedge[yP_kz]\to\neg[xP_kz])$ corresponds to the case in which the primitive kth relation is an equivalence relation. Neither formula is deterministic nor degenerate.

Since we are interested in the formation of *preferences*, we investigate the set of formulae that always define *transitive* relations. We require that the definition of the defined relation be such that, when combined with the requirements on the primitive relation as expressed in the formula $T=\wedge_{k=1,\dots,K}T_k$, transitivity is a *tautology*.

Definition: The formula φ (containing the variables $xP_1y,\dots,xP_Ky$) satisfies *transitivity given $(T_k)_{k=1,\dots,K}$* if the formula $[\varphi(x,y)\wedge\varphi(y,z)\wedge T]\to\varphi(x,z)$ is a tautology.

(The formulae $\varphi(x,z)$ and $\varphi(y,z)$ are obtained from $\varphi(x,y)=\varphi$ after substituting each xP_ky with xP_kz and yP_kz, respectively.)

The requirement that the formula $\psi=[\varphi(x,y)\wedge\varphi(y,z)\wedge T]$ $\to\varphi(x,z)$ be a tautology is at the core of the analysis. The transitivity of the defined relation is required for *all* truth configurations of $\{[xP_ky],[yP_kz],[xP_kz]\}_k$ as long as they satisfy the properties required by $(T_k)_k$.

Note that if we do not impose any restrictions on the basic relations (T_k is "empty"), ψ is reduced to the formula $[\varphi(x,y)\wedge\varphi(y,z)]\to\varphi(x,z)$. However, the set of atomic propositions $\{[xP_ky],[yP_kz]\}_k$, which appear in the conjunction $\varphi(x,y)\wedge\varphi(y,z)$, is disjoint from the set of atomic propositions $\{[xP_kz]\}_k$ appearing in $\varphi(x,z)$. This would make it impossible for $\varphi(x,y)\wedge\varphi(y,z)\to\varphi(x,z)$ to be a tautology unless φ is always true (a tautology) or always false (a contradiction).

Of course, for a given profile of basic relations satisfying the requirements expressed by T, the formula $\psi=[\varphi(x,y)\wedge\varphi(y,z)\wedge T]\to\varphi(x,z)$ may be satisfied for all triples of elements even if ψ is not a tautology. This may occur if the profile of basic relations is not rich enough and configurations of the basic relations that do not satisfy ψ are not realized by any triple of alternatives.

We can now present the main claim of this section:

Claim 4.1: Assume that each T_k is neither *degenerate* nor *deterministic*. If φ is a formula that satisfies transitivity given T, then there is a set $\kappa^* \subseteq \{1, \ldots, K\}$ and a vector of coefficients $\{\delta_k\}_{k \in \kappa^*}$, such that $\varphi \leftrightarrow \bigwedge_{k \in \kappa^*}\delta_k[xP_ky]$ is a tautology (and for every $k \in \kappa^*$, $\delta_k[xP_ky] \wedge \delta_k[yP_kz] \to \delta_k[xP_kz]$ is implied by T_k).

Claim 4.1 states that any definable aggregation rule satisfying transitivity (given T), can be defined by a very simple formula which is essentially a conjunction of primitive propositions or their negations.

Example: Assume that the basic binary relations are both *transitive* and *negatively transitive* (for all k, $T_k = ([xP_ky] \wedge [yP_kz] \to [xP_kz]) \wedge (\neg[xP_ky] \wedge \neg[yP_kz] \to \neg[xP_kz])$. This includes the case in which all primitive relations are linear orderings. Then, any formula of the type $\bigwedge_{k \in \kappa^*}\delta_k[xP_ky]$ for some set K^* satisfies transitivity given T. By the above claim, any formula that satisfies transitivity given T is logically equivalent to a formula of the type $\bigwedge_{k \in \kappa^*}\delta_k[xP_ky]$. These formulae resemble the oligarchic social choice rules in the social choice theory literature.

Example: Consider the operation of constructing a classification system for a set of objects (say, flowers) on the basis of primitive equivalence relations (such as the number of leaves, color, and size). The formula T_k, which embodies the assumption that the kth primitive relation is an equivalence relation, is nondeterministic ($\neg[xP_ky] \wedge \neg[yP_kz]$ does not imply either $[xP_kz]$ or $\neg[xP_kz]$) and is not degenerate. Any formula $\bigwedge_{k \in \kappa^*}[xP_ky]$ for some set κ^* satisfies transitivity and by claim 4.1, every formula in our language that satisfies transitivity given T is logically equivalent to such a formula.

The proof of claim 4.1 (see appendix 1, p. 68) is similar to a standard result in the logic literature called the Craig Lemma (see, for example, Boolos and Jeffrey, 1989):

Craig Lemma: If the formula $\varphi \to \psi$ is a tautology, then there must be a formula, λ, which uses only atomic propositions appearing in *both* φ and ψ such that both $\varphi \rightarrow \lambda$ and $\lambda \rightarrow \psi$ are tautologies.

Note that in our analysis, the preference of a over b relies only on $(aR_k b)_{k=1,\ldots,K}$. This means that the aggregation procedure cannot distinguish between the two cases "$aR_k b$ and $bR_k a$" and "$aR_k b$ and not $bR_k a$." This would be acceptable if R_k was interpreted as a "strong preference of a over b" but not in the case of a weak preference.

To allow for "indifferences" the set of atomic propositions should be expanded to include also the atomic propositions $(yP_k x)_{k=1,\ldots,K}$. The formulae T_k would be required to be a conjunction of $\neg(xP_k y \wedge yP_k x)$ and eight formulae of the type $xP_k y \wedge yP_k z \rightarrow xP_k z$, $xP_k y \wedge yI_k z \rightarrow xP_k z$, where $xI_k y$ is shorthand for $\neg(xP_k y \vee yP_k x)$. By a proof similar to that of claim 4.1 one can show that every definable aggregation rule (in the language of the calculus of propositions with the proposition variables $xP_1 y, \ldots, xP_K y, yP_1 x, \ldots, yP_K x$) that satisfies transitivity and completeness given T is *lexicographic*: In other words, there is an enumeration of $\{1, \ldots, K\}$, $i(1), \ldots, i(K)$, and a vector of coefficients $\{\delta_{i(k)}\}_{k=1,\ldots,K}$ such that the aggregation rule is defined by the formula $\vee_{k=1,\ldots,K}[\wedge_{t=1,\ldots,k-1} xI_{i(t)} y \wedge \delta_{i(k)} xP_{i(k)} y]$. Thus, the K criteria are examined one by one, according to some order of priority. One element, x, is preferred to another element, y, if the first k for which the $i(k)$th criterion is decisive determines that x is preferred. Thus, although economists almost always exclude lexicographic preferences from their models, from the point of view of this analysis they are probably the most natural definable preferences.

4.3 A girl looks for a boyfriend in a foreign town

This section analyzes a second example illustrating the definability condition. Consider the "tale" of a girl who arrives in a foreign town and wishes to find a boyfriend. She obtains information on each candidate in the form of a list of his dates during the past T days. She has yet to meet either the boys or the girls in town (i.e. they can be considered as "unknowns"). All she can glean from any two given names on the list is whether they are identical or not. An alternative and probably better interpretation is of an entrepreneur who is comparing candidates for a job, each of whom provides a list of his past employers about whom the entrepreneur knows nothing.

Following are several principles which the girl may use when she compares two candidates, A and B (A and B are variables, not individual names):

1. Prefer boy A if he dated Alice longer than B had.
2. Prefer boy A if he has dated all the girls in town and B has not.
3. Prefer boy A if all the girls he dated, from the second date on, had previously been dated by B.
4. If the two boys dated the same girl on the first date, prefer the one who dated her longer.
5. Prefer the boy who made the most "switches."
6. Prefer the boy who has dated more girls.

Prior to the formal definitions, let us discuss the definability of these principles. Obviously, these principles are definable in some sense since I have just defined them. However, the question to be analyzed here is whether they are definable in a limited language where: (i) the decision maker can not refer to a boy by his name, (ii) she can refer to a girl only by the term "the girl he has dated at time t" and (iii) the only comparison the decision maker can make between "the girl he has dated at time t" and "the girl he (or someone else) has dated at time s" is whether they are the same girl or not.

Principle (1) requires the decision maker to be able to refer explicitly to the name "Alice." Principle (2) can be stated in the decision maker's limited language using only the equality relation although it requires the use of a quantifier. Principles (3, 4, 5, 6) can be stated in this simple language. However principle (3) does not provide a definition of a preference since (for the case of $T=2$, for example) it implies that the dating history (a,b) is preferred to (b,c), which is preferred to (c,a), which is preferred to (a,b). It is easy to see that principle (4) does not define a preference either. Only principles (5,6) are definable in the limited language and induce preference relations.

We can now move on to the formal analysis. We characterize the binary relations that are definable in the "pure language with equality." This is the language of the calculus of predicates that includes symbols for equality only. In this language, an *atomic formula* is of the type $z_1 = z_2$, where z_1 and z_2 are variable names (such as $x_7 = x_3$ or $x = x_2$). A *formula* is a string of symbols constructed according to the following inductive rules: All atomic formulae are formulae; if φ and ψ are formulae, then $\neg\varphi$, $\varphi \vee \psi$ and $\varphi \wedge \psi$, $\varphi \rightarrow \psi$ and $\varphi \leftrightarrow \psi$ are also formulae; and if x is a free variable (does not appear with a quantifier) in the formula φ, then $\exists x \varphi(x)$ and $\forall x \varphi(x)$ are formulae as well.

A *model* in this simple language is simply a set G, interpreted here as a set of girls. The validity of a formula φ with the free variables $z_1, \ldots, z_K$, in a model G (when we substitute each z_k with an element $a_k \in G$) is defined by the standard truth tables and is denoted by $G \vDash \varphi(a_1, \ldots, a_K)$.

We are interested in preferences on dating profiles of length T that are defined by a formula φ with free variables $x_1, \ldots, x_T, y_1, \ldots, y_T$. Given a set G, the comparison between two boys who have the dating profiles $(a_1, \ldots, a_T)$ and $(b_1, \ldots, b_T)$ is defined by $G \vDash \varphi(a_1, \ldots, a_T, b_1, \ldots, b_T)$. Note that principle (2) can be expressed by the formula $[\forall x \bigvee_{t=1, \ldots, T}(x_t = x)] \wedge [\neg \forall x \bigvee_{t=1, \ldots, T}(y_t = x)]$ while principle (3) can be stated by the formula $\bigwedge_{t=2, \ldots, T} \bigvee_{m=1, \ldots, t-1}(x_t = y_m)$.

In the rest of the section we are interested in formulae that induce binary relations that are transitive and asymmetric under all possible circumstances – i.e. in all models.

Definition: We say that a formula φ with $2T$ free variables induces a transitive and asymmetric relation ordering if the formulae

$$\forall x_1, \ldots, x_T, y_1, \ldots, y_T, z_1, \ldots, z_T [\varphi(x_1, \ldots, x_T, y_1, \ldots, y_T)$$
$$\wedge \varphi(y_1, \ldots, y_T, z_1, \ldots, z_T) \rightarrow \varphi(x_1, \ldots, x_T, z_1, \ldots, z_T)]$$

and

$$\forall x_1, \ldots, x_T, y_1, \ldots, y_T [\varphi(x_1, \ldots, x_T, y_1, \ldots, y_T) \rightarrow \neg\varphi(y_1, \ldots, y_T, x_1, \ldots, x_T)]$$

are tautologies – i.e. they are satisfied in all models.

The analysis of this definability condition consists of three points: *First*, we can assume that the formula φ is written *without quantifiers*. The "pure language with equality" has the property (see, for example, Robinson, 1963) that for every model G (that is, for any set), every formula has a logically equivalent formula containing the same set of free variables with no quantifiers. To illustrate this point, consider the formula

$$\varphi(x_1,\dots,x_T,y_1,\dots,y_T) = [\forall x \bigvee_{t=1,\dots,T}(x_t=x)] \wedge [\neg \forall x \bigvee_{t=1,\dots,T}(y_t=x)],$$

which states that a boy who has dated all the girls in town is preferred to one who has not. Although the quantifier "for all" is used here, the validity of $\varphi(a_1,\dots,a_T,b_1,\dots,b_T)$ in a model G depends only on the number of elements in G and the number of elements in each of the two vectors. If G includes $L \leq T$ elements then, when substituting $(a_1,\dots,a_T)$ for $(x_1,\dots,x_T)$, the validity of the formula $[\forall x \bigvee_{t=1,\dots,T}(x_t=x)]$ is the same as that of a formula stating that in the vector $(x_1,\dots,x_T)$ there are L different elements, which can be written without quantifiers. (Take the disjunction of conjunctions, each of which corresponds to a partition of $\{1,\dots,T\}$ into L nonempty sets $\{e_1,\dots,e_L\}$ and state that $x_i=x_j$ for any i and j that are in the same e_k and $\neg x_i=x_j$, if i and j are in distinct cells of the partition.)

The fact that for any model G, φ has an equivalent formula with no quantifiers implies that for any given model G, φ is equivalent to a disjunction of configurations of the variables $\{x_1,\dots,x_T,y_1,\dots,y_T\}$, where a *configuration* of $\{x_1,\dots,x_T,y_1,\dots,y_T\}$ is the "description of which variables are equal." (In other words, it is a conjunction of formulae where, for each pair of variables z_1 and z_2 in $\{x_1,\dots,x_T,y_1,\dots,y_T\}$, either $z_1=z_2$ or $\neg z_1=z_2$ appears in the conjunction with the following constraint: If $z_1=z_2$ and $z_2=z_3$ appear in the conjunction, $z_1=z_3$ is a conjunct as well.)

Second, for any given cardinality of G, the only preferences that are definable (in this simple language) are those that are induced by preferences on the "dating stability profile" of each candidate. The term "dating stability profile" refers to a partition of $\{1,\dots,T\}$, in which i and j are in the same element of the partition if and only if the boy has dated the same girl at time i and time j. In other words, given a vector $(a_1, ..., a_T)$, define $E(a_1, ..., a_T)$ to be the partition of $\{1, ..., T\}$ in which i and j are in the same partition if and only if $a_i = a_j$. For example, if a boy is fickle and dates a girl for only one period, the corresponding dating structure is the partition $\{\{1\}, ..., \{T\}\}$; a "faithful" boy is characterized by the dating structure $\{\{1, 2, ..., T\}\}$. One can show (see appendix 2, p. 69) that if φ defines an ordering in a model G, then for any two dating profiles with $E(a_1, ..., a_T) = E(b_1, ..., b_T)$, it is not true that

$$G \vDash \varphi(a_1, ..., a_T, b_1, ..., b_T).$$

In other words, the decision maker is indifferent between the two profiles $(a_1, ..., a_T)$ and $(b_1, ..., b_T)$.

We are still left with the possible dependency of the defined preference on the cardinality of the set G: for any integer n there is a formula $\varphi_n(x_1, ..., x_T, y_1, ..., y_T)$ with no quantifiers such that $G \vDash \forall x_1, ..., x_T, y_1, ..., y_T$ $[\varphi_n(x_1, ..., x_T, y_1, ..., y_T) \leftrightarrow \varphi(y_1, ..., y_T, x_1, ..., x_T)]$ for all models G with cardinality n. In other words, the decision maker's preference relation on dating profiles may depend on the number of elements in G. This brings me to the *third* and last point: By the compactness argument there is some n^* such that if G's cardinality is at least n^*, $\varphi(x_1, ..., x_T, y_1, ..., y_T)$ is equivalent to $\varphi_{n^*}(x_1, ..., x_T, y_1, ..., y_T)$.

Summarizing:

Claim 4.2: If $\varphi(x_1, ..., x_T, y_1, ..., y_T)$ is a formula which induces a transitive and asymmetric binary relation, then for any n, there is an ordering $>_n$ on the set of "dating structures" of $\{1, ..., T\}$ such that if the size of G is n,

$$(a_1, ..., a_T) \text{ relates to } (b_1, ..., b_T) \text{ if and only if}$$
$$E(a_1, ..., a_T) >_n E(b_1, ..., b_T)$$

and there exists a number n^* such that all $>_n$ are identical for n greater than some n^*.

Thus, for any set G, the comparison of any two dating profiles of a fixed length T (i) does not depend on any comparison of the girls dated by the two boys (such as $x_3 = y_7$),

(ii) for any given size of G, the comparison depends only on the comparison between the dating stability structures of the two boys and (iii) is constant for sufficiently large sets.

4.4 Conclusion

This chapter has presented two examples of the application of a definability constraint on the set of admissible preference relations. It would also be interesting to investigate the definability-type constraints on the set of *situations* that the decision makers may confront. It is appealing, in my opinion, to investigate models in which the emerging choice problems are limited to those which are defined naturally. The language in which the choice problem is defined may affect the language of the decision maker. Imposing constraints both on the class of situations as well as on the individuals' preference relations, especially in interactive situations, may yield interesting results. This would be a very challenging project, but the proof is, as always, in the pudding. The aim of this chapter, however, is much more modest – to persuade the reader that the definability assumption on preferences is natural as well as analytically tractable.

APPENDIX 1 PROOF OF CLAIM 4.1

A basic proposition in the calculus of propositions states that any formula φ with the atomic propositions $[xP_1y], \ldots, [xP_Ky]$ is logically equivalent to its disjunctive normal form, which is a disjunction $\leftrightarrow_{m \in M} \varphi_m$, where each φ_m is a *truth configuration* of the atomic propositions $[xP_1y], \ldots, [xP_Ky]$ – that is, a formula of the form $\wedge_{k=1,\ldots,K} \delta_k[xP_ky]$, where $\delta_k \in \{-1,+1\}$. We will say that a formula of the type $\wedge_{k \in K^*} \delta_k[xP_ky]$ is a decisive formula if $\wedge_{k \in K^*} \delta_k[xP_ky] \rightarrow \varphi(x,y)$ is a tautology. Let K^* be a minimal (in the inclusion order) subset of $\{1, \ldots, K\}$ for which a decisive formula $\wedge_{k \in K^*} \delta_k[xP_ky]$ exists. It is impossible that two distinct sets $K(1)$ and $K(2)$ are such minimal subsets. If they were, then $(\wedge_{k \in K(1)} \delta_k[xP_ky]) \wedge (\wedge_{k \in K(2)} \delta_k'[yP_kz]) \wedge T \rightarrow \varphi(x,z)$ would be a tautology. By the nondegeneracy condition the formula $(\wedge_{k \in K(1)} \delta_k[xP_ky]) \wedge (\wedge_{k \in K(2)} \delta_k'[yP_kz]) \wedge T$ implies a formula $(\wedge_{k \in K} \varepsilon_k[xP_kz])$ for which $(\wedge_{k \in K} \varepsilon_k[xP_kz]) \rightarrow \varphi(x,z)$ is a tautology, where for all $k \in K$ the formula $\delta_k[xP_ky]\wedge\delta_k'[yP_kz]\rightarrow\varepsilon_k[xP_kz]$ is in T and $K \subseteq K(1)\cap K(2)$. It follows that there is a unique minimal set K^* for which $\wedge_{k\in K^*}\delta_k[xP_ky]\rightarrow\varphi(x,y)$ is a tautology.

Now, assume that there are two tautologies: $\wedge_{k\in K^*}\delta_k[xP_ky]\rightarrow\varphi(x,y)$ and $\wedge_{k\in K^*}\delta_k'[xP_ky]\rightarrow\varphi(x,y)$. Determine $m \in K^*$ such that $\delta_m \neq \delta'_m$. Since T_m is not deterministic, at least one of the four conjunctions $\delta_m[xP_my]\wedge\delta_m[yP_mz]$, $\delta_m[xP_my]\wedge$ $\delta'_m[yP_mz]$, $\delta'_m[xP_my]\wedge\delta_m[yP_mz]$ and $\delta'_m[xP_my]\wedge\delta'_m[yP_mz]$ does not imply the truth value of $[xP_mz]$. If, for example, T_m does not imply any formula of the type $\delta_m[xP_my]\wedge\delta_m[yP_kz]\rightarrow\delta[xP_kz]$ then consider the tautology $[(\wedge_{k\in K^*}\delta_k[xP_ky])\wedge(\wedge_{k\in K^*}\delta_k[yP_kz])\wedge$ $(\wedge_k T_k)]\rightarrow\varphi(x,z)$. Again, it follows that there is a tautology of the type $\wedge_{k\in K^*-\{m\}}\delta_k[xP_kz]\rightarrow\varphi(x,z)$, in contradiction to the minimality of the set K^*.

APPENDIX 2 DATING PROFILES

We will show that for any dating structure, if two boys' dating profiles have this structure, the defined preference cannot prefer one to the other. For notational simplicity, we consider the case in which each boy dates T distinct girls.

For any configuration $\psi(x_1, ..., x_T, y_1, ..., y_T)$ in the disjunctive normal form of φ, where $x_i \neq x_j$ and $y_i \neq y_j$ for all $i \neq j$, denote $N(\psi) = \{t \mid$ there is an s such that $x_t = y_s$ appears in $\psi\}$. Let ψ^* be such a configuration, with the lowest number of elements in $N(\psi)$. It is impossible that $N(\psi^*) = 0$, since ψ^* would then be the conjunction of all formulae of the type $\neg z_1 = z_2$ for all $z_1 \neq z_2 \in \{x_1, ..., x_K, y_1, ..., y_K\}$ and the asymmetry tautology would not hold.

It follows that

$$\forall x_1, ..., x_T, y_1, ..., y_T[\psi^*(x_1, ..., x_T, y_1, ..., y_T)\wedge\psi^*(y_1, ..., y_T, z_1, ..., z_T)$$
$$\rightarrow\varphi(x_1, ..., x_T, z_1, ..., z_T)]$$

is a tautology. Let ψ be a configuration which satisfies that for any t and r either $x_t = y_r$ or $\neg x_t = y_r$ appears within the configuration and $x_t = y_r$ appears in the conjunction if and only if there is an s such that both $x_t = y_s$ and $x_s = y_r$ appear in ψ^*. Then ψ must be a configuration in the disjunctive normal form of $\varphi(x_1, ..., x_T, y_1, ..., y_T)$. $N(\psi)$ is a subset of $N(\psi^*)$ and thus, by the minimality of $N(\psi^*)$ it must be the case that $N(\psi^*) = N(\psi)$. Denote by σ the permutation on the set $N(\psi^*)$, defined by $\sigma(i) = j$ if $x_i = y_j$ appears in ψ^*. There must be an integer n such that σ^n is the identical permutation. The formula φ is the disjunction of conjunctions and one of these must be the conjunction of $x_i = y_i$ for $i \in N(\psi^*)$ and the negation of all other equalities, contradicting the asymmetry tautology.

REFERENCES

Boolos, G.S. and R.C. Jeffrey (1989) *Computability and Logic*, Boston and London: Cambridge University Press

Crossley, J.N., J.C. Stillwell, C.J. Brickhill, N.H. Williams and C.J. Ash (1990), *What Is Mathematical Logic?* Oxford: Dover Publications

Robinson, A. (1963) *Introduction to Model Theory and to the Meta Mathematics of Algebra*, Amsterdam: North Holland

Rubinstein, A. (1978) "Definable Preference Relations – Three Examples," Center of Research in Mathematical Economics and Game Theory, RM 31, the Hebrew University of Jerusalem

(1984) "The Single Profile Analogues to Multi-profile Theorems: Mathematical Logic's Approach," *International Economic Review*, 25, 719–30

(1998a) *Modeling Bounded Rationality*, Cambridge, MA: MIT Press

(1998b) "Definable Preferences: An Example," *European Economic Review*, 42, 553–60

(1999) "Definable Preferences: Another Example (Searching for a Boyfriend in a Foreign Town)," *Proceedings of the 11th International Congress of Logic, Methodology and Philosophy*, forthcoming

Shelly, M.Y. and G.L. Bryan (1964) "Judgments and the Language of Decisions," in Shelly, M.Y. and G.L. Bryan (eds.), *Human Judgment and Optimality*, New York: John Wiley

CHAPTER 5

ON THE RHETORIC OF
GAME THEORY

5.1 Introduction

This chapter contains some comments on the language used by game theorists. As such, it is quite different from the rest of the book. Why am I interested in the study of the rhetoric of Game Theory? Admittedly, "I am not thrilled" with the fact that Game Theory is viewed by many as "useful" in the sense of providing a guide for behavior in strategic situations. This view is reinforced by the language we game theorists use.

Consider, for example, John McMillan's *Games, Strategies and Managers* (1992). On the cover of the paperback edition is a quote from *Fortune* praising the book as "the most user friendly guide for business people," and Akira Omori of Arthur Andersen Co. is quoted as saying that "[the book] will be helpful both for beginning managers and for advanced strategic planners. In fact, I would recommend that anyone engaged in the US–Japan negotiations read the book."

Or consider Avinash Dixit and Barry Nalebuff's bestseller, *Thinking Strategically*. The *Financial Times* (December 7, 1991) says of the book: "*Thinking Strategically* ... offers essential training in making choices and weighing possibilities not only in business but in daily life." Schlomo Maital says in a review of the book published in *Across the Board* (June 1991): "Ever since 1926, when John von Neumann, a brilliant Hungarian mathematician and physicist, published

Many of the ideas in this chapter have appeared in my previous writings. In particular, the discussion of "strategy" in section 5.3 is based on Rubinstein (1991).

his path-breaking paper on Game Theory, it has been known that analytical models of games could be built. These models make it possible to find the best strategies and the best solutions for games, including ones that embody complex business dilemmas."

The public misconception of these books is reinforced by the authors themselves. McMillan says: "Stripped of its mathematics and jargon, it can be useful to people in managerial situations faced with important decisions and strategy," and the opening sentence of his book asks: "What is Game Theory about, and how can it be used in making decisions?" Dixit and Nalebuff write: "It is better to be a good strategist than a bad one, and this book aims to help you improve your skills at discovering and using effective strategies," "Our aim is to improve your strategy IQ," and "Our premise in writing this book is that readers from a variety of backgrounds can become better strategists if they know these principles."

The phenomenal success of Game Theory and the perception that it may "improve strategy IQ" are linked, in my opinion, to the rhetoric of Game Theory. I doubt that Game Theory would have attracted the same attention if it were named "the theory of interactions between rational agents." The fact that one might conceivably find a book "about" Game Theory on the *New York Times* bestseller list and that students are fascinated by the "Prisoner's Dilemma" stems from the natural associations of game theoretic jargon: Words like "game," "strategy," and "solution" stimulate our imagination. Is the use of these words justified?

Words are a crucial part of any economic model. An economic model differs substantially from a purely mathematical model in that it is a *combination* of mathematical structure and *interpretation*. The names of the mathematical objects are an *integral* part of an economic model. When mathematicians use everyday concepts such as "group" or "ring," it is only for the sake of convenience. When they name a collection of sets a "filter," they do it in an associative manner and in principle they could call it an "ice cream cone." When they use the term "good ordering" they are not assigning any positive ethical value. The situation is somewhat different in mathematical logic in which, for example, the names of the connectives are strongly related to their common meaning. In economic theory, and Game Theory in particular, interpretation is an essential ingredient of any model. A game varies according to whether the players are human beings, bees, or different "selves" of the same person. A strategic game changes entirely when payoffs are switched from utility numbers representing von Neumann and Morgenstern preferences to sums of money or to measures of evolutionary fitness.

The evaluation of Game Theory rhetoric is also important from the public's point of view. The public's interest in Game Theory is at least partially the result of efforts on the part of academic economists to emphasize the practical value of Game Theory as, for example, a guide to policy makers. Consultants use the professional language of Game Theory in their arguments and it is in the public interest that these arguments be properly understood so that the rhetoric does not distort their real content. Although the discussion of the usefulness of Game Theory can take place on an abstract level without referring to the substance of game theoretic models, it does require an understanding of Game Theory beyond that required for the "Prisoner's Dilemma."

I am doubtful about the practical applicability of Game Theory. However, I do not feel pessimistic since I don't regard applicability as necessarily a virtue. I do not believe that the study of formal logic can help people become "more logical," and I am not aware of any evidence showing that the study of probability theory significantly improves people's ability to think in probabilistic terms. Game Theory is related to logic and probability theory and I doubt that it could prove useful to negotiators or other types of players. In fact, I suspect that Game Theory could even be misleading them since people often ignore the subtleties of game-theoretic arguments and treat solution concepts as instructions. (Nevertheless, I feel obliged to mention that I failed to demonstrate this in an experiment carried out in collaboration with a group of Tel Aviv

University students. The results of the experiment were inconclusive.)

In fact, McMillan's book is full of reservations on the direct usefulness of Game Theory: "Game Theory does not purport to tell managers how to run their business"; "Game Theory does not eliminate the need for the knowledge and intuition acquired through long experience"; "Game Theory offers a short cut to understanding the principles of strategic decision making. Skilled and experienced managers understand these principles intuitively, but not necessarily in such a way that they can communicate their understanding to others"; and "Game Theory, then, is a limited but powerful aid to understanding strategic interactions." Dixit and Nalebuff (in spite of their statement "Don't compete without it") state that "in some ways strategic thinking remains an art."

And indeed when it comes to describing the usefulness of Game Theory in actual cases, all authors adopt more modest goals and become rather vague. Despite the promise to be a user-friendly guide, McMillan concludes his chapter on "Negotiating," a topic which he considers to be the "archetypical problem of Game Theory," with the modest statement: "What advice for negotiators does Game Theory generate? The most important idea we have learned in this chapter is the value of putting yourself in the other person's shoes and looking several moves ahead." Dixit and Nalebuff's "Rule 1" is to "Look ahead and reason back." They state the general principle for sequential-move games to be "that each player should figure out the other players' future responses, and use them in calculating his own best current move," and the final sentence of their book reads "You can bet that the most heavily played machines are not the ones with the highest payback."

If the most important lesson from reading a "user-friendly guide to Game Theory" is the value of putting yourself in the other person's shoes, and if improving your strategic IQ is boiled down to "anticipating your rival's response," I am not convinced that Game Theory is more valuable than a detective novel, a romantic poem, or a game of chess in achieving these goals.

Incidentally, is it clear that academic knowledge is necessarily valuable? Is it clear that improving strategic IQ is a desirable goal? David Walsh, a *Washington Post* writer, said "The problem is, of course, that if Dixit and Nalebuff can improve your strategic IQ, they can improve your competitor's as well." This hints at an important point which is valid for other social sciences as well: Improving strategic reasoning, even if it is possible (and I doubt that it is), should be evaluated by social standards which take into account not only the success of a company's strategic analyst but general welfare as well. Is it desirable to improve the strategic IQ of a small sector of the population, especially one whose position we might not wish to improve?

Incidentally, the discussion of the rhetoric of economics is not a new subject. The best known research on this subject was published in the mid-1980s (McCloskey, 1998, 2nd edn.) and was followed by an extensive literature (see, for example, Henderson, Dudley-Evans and Backhouse, 1993). My contribution in this chapter is quite limited. I will point out some of the rhetorical problems associated with the language of Game Theory. I will argue that the rhetoric of Game Theory is actually misleading in that it creates an impression that it has more practical application than it actually does. In particular, I wish to show that:

(1) People look to Game Theory for advice on what strategy to adopt in game-like situations; however, the basic notion of "strategy" in Game Theory can hardly be interpreted as a course of action.
(2) The use of formulae in Game Theory creates an illusion of preciseness which does not have any basis in reality.

5.2 Strategy

"Strategy" in daily language

One of the central terms in Game Theory is "strategy." What is a strategy in ordinary language? McMillan (1992) teaches us that the source for the word "strategy" is the Greek word for the leader of an army, which is rather different from its usage today. Webster's dictionary defines the word as "a method for making or doing something or attaining an end," and the *Oxford English Dictionary* defines the word as "a general plan of action." We talk about the strategy to win a war, the strategy of managers to make profits or avoid being fired, and the strategy to survive a hurricane.

The informal definition of strategy according to game theorists is not far from the one in everyday language. Martin Shubik refers to a strategy as "a complete description of how a player intends to play a game, from beginning to end." Jim Friedman defines it as "a set of instructions." And John McMillan defines strategy as "a specification of actions covering all possible eventualities." Thus, it is appropriate to ask whether the formal definition of the term "strategy" in Game Theory is similar to its everyday meaning. If it is not, then Game Theory's usefulness is put into serious doubt; and even if it is, there is still a long way to go before we can be persuaded that game-theoretic results can teach something to managers.

A strategy in an extensive game

In an extensive game, a player's strategy is required to specify an action for each node in the game tree at which the player has to move. Accordingly, a player has to specify an action for every sequence of events which is consistent with the rules of the game. As an illustration, consider the two-player game form in figure 5.1:

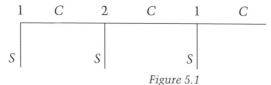

Figure 5.1

According to the natural definition of strategy as a "complete plan of action," player 1 is required to specify his behavior, "Continue" (*C*) or "Stop" (*S*), at the initial node and, if he plans to "Continue," to make provisional plans for his second decision node in the event that player 2 chooses *C*. In contrast, the game-theoretic definition of strategy requires player 1 to specify his action at the second decision node, even if he plans to "Stop" the game at the first node. Here, as in any game which requires a player to make at least two consecutive moves (and most of the games which have been analyzed in economic theory fall into this category), a strategy must specify the player's actions even after histories which are inconsistent with his own strategy.

Why does the notion of strategy used by game theorists differ from a "plan of action"? If we were investigating only Nash equilibria of extensive games, then the game-theoretic definition would indeed be unnecessarily broad. The broad definition is, however, necessary for testing the rationality of a player's plan, both at the beginning of the game and at the point where he must consider the possibility of responding to an opponent's potential deviation (the subgame perfect idea). Returning to the game form above, assume that each player plans to choose "Stop" at his first decision node. Testing the optimality of player 2's plan following player 1's deviation requires that player 2 specify his expectations regarding player 1's plan at his second decision node. The specification of player 1's action after both players have chosen C provides these expectations and has to be interpreted as what would be player 2's (as opposed to player 1's) *belief* regarding player 1's planned future play should player 1 decide to deviate from what was believed to be his original plan of action. Thus, a strategy encompasses not only the player's plan but also his opponents' expectations in the event that he does not follow that plan. Hence, an equilibrium strategy describes a player's plan of action as well as those considerations which support the optimality of his plan (i.e. preconceived ideas concerning the other players' plans) rather than being merely a description of a "plan of action." A profile of strategies provides a full analysis of the situation and not merely a tuple of plans of actions.

A mixed strategy in a strategic game

The naive interpretation of "mixed strategy" requires that a player use a roulette wheel or some other random device to decide upon an action. Normally, we are reluctant to believe that our decisions are made at random; we prefer to believe that there is a reason for each action we take. Most game theorists never considered using random devices as part of a strategy before being exposed to Game Theory. Indeed, the concept of mixed strategy has been a target for criticism. To quote Aumann (1987): "Mixed strategy equilibria have always been intuitively problematic" and Radner and Rosenthal (1982): "One of the reasons why game theoretic ideas have not found more widespread application is that randomization, which plays a major role in Game Theory, seems to have limited appeal in many practical situations."

There are obviously cases in which players do choose random actions. When a principal (for example, an employer or the government) monitors his agents (employees or citizens), often only a few are monitored and they are chosen randomly. Of course, this is not a "mixed strategy." The stochastic rule used for auditing is actually a pure strategy. The principal is not indifferent between monitoring no agents and monitoring all of them. His "pure strategy" is the parameters of the randomization to be used. As an additional example, a parent who must allocate one piece of candy to one of his two children may strictly prefer to flip a coin. However, this flip of a coin is not a "mixed strategy" either. Typically, the parent would not be indifferent between procedures for allocating the candy. An allocation procedure should be regarded as a pure strategy even if it involves flipping a coin.

The game-theoretic literature suggests a few more interpretations of this concept. The *large population frequencies* interpretation of a mixed strategy was used in chapter 2. The game is played occasionally between players, each drawn from a population homogeneous in its interests but heterogeneous in its actions. The mixed strategy describes the distribution of actions taken by the agents who are drawn from the population attached to any given role in the game. By a different interpretation, a mixed strategy is conceived of as a plan of action which is dependent on private information not specified in the model. According to this interpretation, a player's behavior is actually deterministic although it appears to be random. If we add this information structure to the model, the mixed strategy becomes a pure one in which actions depend on extraneous information. This interpretation is problematic since it is not clear why people would base their behavior on factors which are clearly irrelevant to the situation (see Harsanyi, 1973, for a more sophisticated interpretation which responds to this criticism).

There is an additional interpretation which has gained popularity during the past decade. According to this interpretation, a mixed strategy is the belief held by all *other* players concerning a player's actions (see Aumann, 1987). A mixed strategy equilibrium is then an n-tuple of common knowledge expectations; it has the property that, given beliefs, all actions with a strictly positive probability are optimal. Thus, the uncertainties behind the mixed strategy equilibrium are viewed as an expression of the lack of certainty on the part of the other players rather than an intentional plan of the individual player.

Does it matter?

It would appear that two central versions of the game-theoretic notion of a "strategy" – i.e. an extensive form strategy and a mixed strategy in strategic games – are more appropriately interpreted as beliefs rather than plans of actions. However, if strategies are beliefs rather than plans of actions, the entire manner of speaking in Game Theory has to be reassessed. We can speak of player 1 choosing his strategy, but does he choose player 2's belief about his strategy?

As a demonstration of the difficulties which follow from the beliefs interpretation, consider the sequential games literature in which authors sometimes *assume* that strategies are stationary in the sense that a player's behavior is independent of the history of the game. This literature presents stationarity as an assumption related to simplicity of behavior. For example, consider player 1's strategy "always play a" in a repeated game. This strategy is simple in the sense that the player plans to make the same move independently of the other players' actions. However, this strategy also implies that player 2 believes that player 1 will play a even if player 1 has played b in the first 17 periods of the game. Thus, stationarity in sequential games implies not only simplicity but also passivity of beliefs. This is counter-intuitive, especially if we assume simplicity of behavior. If player 2 believes that player 1 is constrained to choose a stationary plan of action, then player 2 probably should believe (after 17 repetitions of the action b) that player 1 will continue to play b. Thus, by assuming passivity of beliefs, we eliminate a great deal of what sequential games are intended to model – i.e. the changing pattern of players' behavior and beliefs as players accumulate experience.

With regard to mixed strategies, the adoption of the beliefs interpretation requires the reassessment of much of applied Game Theory. In particular, it implies that equilibrium does not lead to a statistical prediction of players' behavior. Any action taken by player i as a best response given his expectation about the other players' strategies is consistent as a prediction for i's future action (this might even include actions which are outside the support of the mixed strategy). This renders meaningless any comparative statics or welfare analysis of the mixed strategy equilibrium and brings into question the enormous economic literature which utilizes mixed strategy equilibrium.

5.3 The "numbers" illusion

Although the question of whether we should represent an individual by a preference relation or a utility function may seem immaterial (since a utility function simply represents a preference relation numerically), I will argue that this is not the case and that the "numeration of Game Theory" sends a very misleading message with regard to the precision and relevancy of Game Theory.

To demonstrate this point, let us briefly discuss the Nash bargaining solution which is one of the most fundamental models in modern economic theory. Nash's theory is designed to provide a "prediction" of the bargaining outcome based on two elements: (1) the bargainers' preferences which are defined on the set of possible agreements (including the event of disagreement), and (2) the bargainers' attitude towards risk.

The primitives of Nash's (two-person) *bargaining problem* are the *"feasible set"* and a *"disagreement point"*, denoted by S and d, respectively. In Nash's formalization of the bargaining problem each element of S corresponds to the pair of numbers interpreted as the utility levels obtained by the two bargainers at one (or more) of the possible agreements. The utilities are understood to be von Neumann–Morgenstern in that they are derived from preference relations over lotteries that satisfy the expected utility assumptions. The disagreement point is modeled as one of the points in S. The second basic concept of Nash's bargaining theory is the *bargaining solution*, defined as a function which assigns a unique pair of utility levels to each bargaining problem (S,d). Thus, a bargaining solution is supposed to provide a "unique prediction" of the bargaining outcome (in utility terms) for each of the problems in its domain.

Nash (1950) showed that there is a unique bargaining solution satisfying the following four axioms: *Invariance to Positive Affine Transformations* (the re-scaling of each bargainer's utility values by a positive affine transformation shifts the solution accordingly); *Symmetry* (a symmetric problem has a symmetric solution point); *Pareto Optimality* (for each bargaining problem, the solution is on the Pareto frontier); and *Independence of Irrelevant Alternatives* (IIA) (the shrinking of the set of alternatives without eliminating the solution point and without eliminating the disagreement point does not alter the solution point) which is the most problematic of the axioms in terms of its interpretation. The unique solution satisfying these axioms is the Nash solution – i.e. the function

$$N(S,d) = \mathrm{argmax}\{(u_1 - d_1)(u_2 - d_2) \mid (u_1,u_2) \in S \text{ and } u_i \geq d_i \text{ for both } i\}.$$

If the task of bargaining theory is to provide a "clear-cut" numerical prediction for a wide range of bargaining problems, then Nash certainly achieved this goal. The fact that it is defined by a simple formula is a significant advantage of the theory, especially when we try to embed it in a larger model that contains a bargaining component. But can this prediction be tested as in the sciences? Does the formula predict how people will share a pie with the accuracy that we can calculate the moment that a stone falling from a tower will hit the ground?

I doubt it very much. The significance of Nash's formula derives from its abstract meaning regardless of its testability. The use of numbers, even if analytically convenient, obscures the meaning of the model and creates the illusion that it can produce quantitative results. Does the Nash bargaining solution perhaps have a different, more abstract meaning?

On the face of it, the answer is "no." The formula of the Nash bargaining solution lacks a clear meaning. What is the interpretation of the product of two von Neumann–Morgenstern utility numbers? What is the meaning of the maximization of that product? Can we consider the maximization of the product of utilities to be a "useful" principle for resolving conflicts?

The use of numbers to specify the bargaining problem has obscured the meaning of the Nash bargaining solution. Were game theorists to use a more natural language to specify the model, the solution would become clearer and more meaningful. In Rubinstein, Safra and Thomson (1992) the indirect language of utility is replaced by the direct language of alternatives-preferences. The model's elements are interpreted directly using the words "alternative," "lottery," "disagreement" and "preference." The Nash problem, (S, d), is "replaced" by a four-tuple $(X, D, \geq_1, \geq_2)$ where X is a set of feasible agreements, D is a "disagreement" event, and $(\geq_1, \geq_2)$ is the bargainers' preference relations. As Nash's theory aims to predict the bargaining outcome as a function of the players' interests and their attitude to risk, preferences are taken to be defined over the set of lotteries whose "certain prizes" are the agreements in X and the disagreement event D.

Using this model, an alternative definition of the Nash bargaining solution can be formulated:

On the rhetoric of game theory

An *(ordinal) Nash solution* for the problem $(X,D, \geq_1, \geq_2)$ is an alternative y^* such that for all $p \in [0,1]$ and all $x \in X$ and i, if $p^\circ x \geq_i y^*$ then $p^\circ y^* \geq_i x$ (where $p^\circ x$ is the lottery which gives x with probability p and D with probability $1-p$).

The suggested interpretation of this solution is as follows: A solution is a "convention" which attaches a unique agreement to any bargaining problem. The solution embodies the assumption that players are aware that when they raise an objection to an alternative, they risk that the negotiations will end in disagreement. The Nash solution agreement is the only one which satisfies the following property:

If–

- It is worthwhile for one of the players to demand an improvement in the convention, even at the risk of a breakdown in negotiations,

then–

- It is optimal for the other player to insist on following the *convention* even at the risk of a breakdown in negotiations.

In other words, the Nash bargaining solution is an agreement satisfying the condition that any argument of the type "You should agree to my request x without delay, since this is preferable to you over the convention y^* given the probability $1-p$ of breakdown" is not profitable for the player making the request when he takes into account the same probability of breakdown.

The Gulf War provides a concrete example of this definition: The bargainers were Iraq and the USA. The set of agreements contained the various possible partitions of the land in that region. The disagreement event was a war. When Saddam Hussein moved his troops he deliberately took the chance that the situation would deteriorate into war before the USA capitulated. Apparently he preferred the risk of war with a certain probability while hoping that the USA would agree to his request to annex Kuwait. In contrast to his expectation, the USA preferred to take the risk of war and basically demanded a return to the status quo rather than give in to Iraq's demands. If the USA had yielded to Iraq, it would have meant that the pre-invasion borders were not part of a Nash bargaining outcome.

When preferences satisfy the expected utility assumptions, this definition and the standard one converge (see Rubinstein, Safra and Thomson, 1992). However, a byproduct of the ordinal definition is that Nash's theory can be extended to a domain of preferences beyond those which satisfy the expected utility theory. The language of preferences does not require the preferences to satisfy the expected utility axioms; thus the definition can be applied to a wider set of preferences on the set of lotteries over the agreements and the disagreement point. One can show that for a wider class of preferences (which includes all preferences satisfy the expected utility axioms), the Nash solution is well defined and is the unique solution which satisfies a set of axioms which resemble the Pareto, Symmetry, and IIA axioms.

Indeed the switch to the alternatives-preferences language requires a restatement of the entire Nash theory. The details are beyond the scope of this book and I will comment only on the change in the meaning of the IIA axiom. The original IIA axiom requires that if u^* is a utility pair which is the solution of the problem (T,d), and if u^* is a member of a set S which is a subset of T, then u^* is also the solution outcome of (S,d). As has often been emphasized, this justification of the IIA axiom suits a normative theory in which the solution concept is intended to reflect the social desirability of an alternative. When bargaining is viewed as a strategic interaction of self-interested bargainers, the validity of the IIA axiom is questionable.

The following is an alternative statement of the IIA axiom which does not require a comparison between problems having different sets of alternatives. The "ordinal-IIA" states that a solution to a bargaining problem y^* remains invariant to a change in i's preference that would make him less willing to raise objections to y^*. Formally, assume that y^* is the solution to the problem $(X,$

$D, \geq_1, \geq_2$) and let $\geq'_i$ be a preference which agrees with $\geq_i$ on the set of deterministic agreements X such that:

(i) for all x such that $x \geq_i y^*$, if $p^o x \sim_i y^*$ then $p^o x \leq'_i y^*$ and

(ii) for all x such that $x \leq_i y^*$, if $x \sim'_i p^o y$ then $p^o x \leq'_i y^*$.

Hence, y^* is also the solution to the problem $(X, D, \geq'_i, \geq_j)$.

The switch of player i's preference from $\geq_i$ to $\geq'_i$ reflects his increased aversion to the risk of demanding alternatives which are better for him than the outcome y^*. The change in player i's preference makes player i "less willing" to object. Though bargainer i, with preference $\geq'_i$, has the same ordinal preference over X as $\geq_i$, he is less willing to take the risk of demanding agreements which are better than x^*. The axiom states that this change does not affect the bargaining outcome. In other words, changing player i's preference to make him less willing to achieve agreements which are better than y^* does not change the bargaining outcome. With this interpretation of IIA, the link between the axiomatization and the ordinal definition of the Nash solution seems more comprehensible. This axiom makes sense when the underlying bargaining procedure is such that there is a clear convention and an objection can be "protected" from other objections only by insisting on the retention of the convention. It makes less sense when counter-arguing can be carried out by other means, such as raising alternative demands, or when agreement is reached by the two parties gradually reducing their demands.

Let us return to the main point of this section: The language of utility allows the use of geometrical presentations and facilitates analysis; in contrast, the numerical presentation results in an unnatural statement of the axioms and the solution. The switch to the language of alternatives-preferences allows a more natural statement and characterization of the Nash solution. One may argue that the equivalence between the terms "preferences" and "utility" in economic models implies equivalence in their use as well. However, as demonstrated here, the choice of language affects the meaning of the results. The use of the language of utility calls for arithmetic operations (such as multiplying utilities) which do not necessarily have meaningful interpretations. The use of the language of preferences leads to the derivation of a solution which resembles a consideration which could conceivably be used in real life. Nevertheless, I am not convinced that Nash's theory has done more than clarify the logic of one consideration which influences bargaining outcomes. I can not see how this consideration will comprehensively explain real-life bargaining results.

5.4 The term "solution"

In the previous two sections I discussed the use of the term "strategy" and the substitution of "utilities" for preferences. In this section, I want to briefly discuss another key term in Game Theory – "solution." The structure of a game-theoretic analysis is as follows: first, the modeler spells out the parameters of the game (identities of the players, rules, information available to the players, and the players' interests). The analysis then moves on to "finding solutions." The result of applying a solution concept to a game is a set of profiles each of which assigns a strategy to each player. A solution concept is a method which assigns to each game a set of profiles of strategies satisfying certain conditions of stability and rationality.

This form of investigation distinguishes between the assumptions embedded in the solution concept and those underlying the description of the game. The former are regarded as being most solid, although they are in fact very strong assumptions concerning the players' behavior and process of deliberation. Since they are buried in the solution concept, these assumptions are almost never discussed by users of Game Theory. Incidentally, these assumptions are not all that clear to game theorists either, and it was a long time before game theorists began examining the epistemological assumptions underlying the various solution concepts.

Thus, whereas applied economists often debate the assumptions of a model they pay almost no attention to the appropriateness of the assumptions which are incorporated into the solution concept. A solution concept is treated as a machine which receives a situation as an input and produces an output which is interpreted as a prediction of the way in which the game will be resolved.

In my opinion, the use of the particular word "solution" is central in creating the view that Game Theory is a "prediction machine." The term "solution" implies the existence of a problem. In the case of Game Theory the problem is to figure out "what will happen" or "what move to make." The word "solution" creates a sense of being clearly defined and correct. The set of solutions for a mathematical equation is certainly well defined; can we say the same of the set of solutions to a game? Though the solution of a game is often viewed as being analogous to the solution of an equation or maximization problem, in fact, a game-theoretic solution concept is no more than the statement of a particular consideration which we have chosen to focus on.

Another term often used during discussions of the meaning of solution concepts is *"recommendation."* It is often said that a solution concept provides a profile of recommendations which are not "self-defeating." For example, a Nash equilibrium is said to be a profile of recommendations, one for each player, with the property that if players believe that the recommendations will be followed, then no one has an interest in not following his recommendation.

I think that the use of the term "recommendation" here is misleading. The term implies "authority," "an instruction," and the "right thing to do." Can the term Nash equilibrium be considered as a synonym for recommendation? A recommendation must be given by someone. Who is making the recommendation in this case? Is he just a game theorist? What are his goals? Can a Pareto-inferior Nash equilibrium be considered a recommendation? And why should the source of the recommendation limit himself to recommendations which are not self-defeating? Would it not be better to give each of the two players in a centipede game a recommendation to never stop the game even though it is not a self-defeating recommendation?

5.5 Final comments

The reader may have gotten the impression that the purpose of this chapter is to diminish the importance of Game Theory and to discourage potential users. On the contrary, I think that Game Theory is moving into new areas of interest and we will soon be witness to exciting developments. However, I have no expectations of Game Theory becoming "practical" as the term is understood by most people. Since I have never understood these expectations of Game Theory I am neither disappointed nor pessimistic.

I see analytical economic theory as a search for *connections* between concepts, assumptions, and assertions which we use in understanding human interaction. Economic theory is an investigation of the different types of considerations used by human beings to analyze human interaction. A strategic situation can be analyzed in many ways. Typically, there are many different arguments supporting one course of action or another. The most we can do analytically (as opposed to empirically) is to clarify the connections between the different types of analysis. "Drawing links" and "understanding," however, stop far short of giving advice and being practical. In my opinion, achieving a better understanding of the arguments used in human interaction and reasoning is in itself a very worthwhile and exciting project. For other views of the meaning of Game Theory, see Aumann (1987) and Binmore (1990).

At a conference where I presented this chapter, I was asked by a listener whether I feel frustrated that my work and views have made no contribution to the world. Perhaps I am. However, I don't think that this lecture is useless in the sense of "influencing the world." There are few tasks in the academic world of the social sciences more important than fighting unjustified authority. This chapter is my modest contribution to this important battle.

CONCLUDING REMARKS

At the opening of this short series of lectures, I briefly presented five issues which fall under the heading of "economics and language." By now you have realized that the issues were indeed quite distinct. I found it difficult to point out the common factor among the topics other than the title and the speaker.

I am deeply grateful to three people who have (surprisingly) agreed to comment on this manuscript. Tilman Börgers, who was also the original discussant in Cambridge in 1996, Bart Lipman, and Johan van Benthem. Tilman and Bart bring some perspectives from economists, and Johan brings his perspective as a logician. It must be hard to write such a review, especially when one wants to criticize the author. Thus, I wish that the reader will pull out from the comments especially the critical comments and the suggestions for further research.

The three discussants helped me out in finding the common features of the first four essays. As Bart Lipman puts it, the first four chapters touch three related questions: "What gives a statement its meaning?," "Why is language the way it is?," and "How does language affect action?"

The discussants also persuaded me that the last chapter is very different in nature. If it belongs to this book beyond my rhetorical use of the phrase "economics and language", it is only because it presents my general approach to economic theory, which is a thread running through these lectures. By this approach economic theory

- speaks more about methods and modeling issues
- talks less about economic substance

Some concluding remarks

- emphasizes the complexity issue as a major element in the discussion
- treats economic theory as the study of the arguments used by human beings

Illuminating the arguments which human beings use in social interactions and investigating "the logic of the situation," a phrase Karl Popper uses to describe the aim of the social sciences, is indeed what I perceive to be the objective of economic theory.

REFERENCES

Aumann, R. (1987) "What is Game Theory Trying to Accomplish?," in K.J. Arrow and S. Honkapohja (eds.), *Frontiers of Economics*, Oxford: Blackwell

Binmore,K. (1990), "Aims and Scope of Game Theory," in K. Binmore, *Essays on the Foundations of Game Theory*, Oxford: Blackwell

Dixit, A.K. and B.J. Nalebuff (1991) *Thinking Strategically*, New York: Norton

Harsanyi, J. (1973) "Games with Randomly Distributed Payoffs: A New Rationale for Mixed-strategy Equilibrium Points," *International Journal of Game Theory*, 2, 486–502

Henderson, W., T. Dudley-Evans and R. Backhouse (eds.) (1993) *Economics and Language*, London: Routledge

McCloskey, D.N. (1998) *The Rhetoric of Economics (Rhetoric of the Human Sciences)*, 2nd edn., Madison: University of Wisconsin Press

McMillan, J. (1992) *Games, Strategies, and Managers*, Oxford: Oxford University Press

Nash, J. (1950) "The Bargaining Problem," *Econometrica*, 18, 155–62.

Radner, R. and R. Rosenthal (1982) "Private Information and Pure Strategy Equilibrium," *Mathematics of Operation Research*, 7, 401–9.

Rubinstein, A., (1991) "Comments on the Interpretation of Game Theory," *Econometrica*, 59, 909–24.

Rubinstein, A., Z. Safra and W. Thomson (1992) "On the Interpretation of the Nash Bargaining Solution," *Econometrica*, 60, 1171–86.

PART 3

COMMENTS

ECONOMICS AND LANGUAGE

Johan van Benthem

1 The ubiquity of language

Language is the air that we breathe when thinking and communicating, often without noticing. It makes human cognition possible, and at the same time constrains it, in often invisible, but very real ways. Philosophy took a "linguistic turn" in this century, when this crucial medium became a focus of attention, out in the open, especially in the analytical tradition. The result were new insights into the structure and functioning of language, which have gone into modern logic and linguistics, often under the heading of "semantics." Ariel Rubinstein makes such a linguistic turn as an economist. He investigates "the economics of language" – i.e. optimization mechanisms that make language work, both as a structure and as an activity, but also "the language of economics," linguistic mechanisms underlying economic behavior and our theorizing about it. More generally, his book is an attempt to position Game Theory in a broad intellectual landscape of reasoning and communication: which I find very congenial. Indeed, the author writes about Paul Grice's views in "Logic and Conversation": "[this] is essentially a description of one agent thinking about how another is thinking. This is precisely the definition of strategic reasoning and is the essence of game theory." I found that quote interesting as a logician, because the analysis of many-agent communication is also a central topic on the modern agenda of logic, computer science, and cognitive science. Are we all in the same boat? Let's compare Rubinstein's agenda with that of people like me.

Attention to language is of the essence in logic. The

logical forms that drive valid reasoning are linguistic in nature, and hence issues of syntax and semantics have been central to the development of logic from the start. True, logicians have found it necessary to design special-purpose formal languages, as a supplement to natural languages, that have started leading lives of their own. For example, when the author wants to survey possible forms of definition, he uses a first-order predicate logic, a nineteenth-century formalism. But such formal languages are also used in the study of various aspects of natural language. Grottlob Frege, the inventor of modern logic, compared its use to that of a microscope, versus natural language, which is like the eye. As he says, formal languages are more precise for specific purposes, but natural language remains the much more versatile general-purpose instrument. Logical theory since Frege has been described as a "marriage of mathematics and linguistics." Logicians will typically study two sides of a coin: different syntactic types of statement are connected with matching semantic kinds of behavior (e.g. invariances across situations), often described mathematically.

2 From description to explanation

Much research in linguistics and logic is content with describing the facts of language and reasoning as they are – and not to explain how they came about, or continue functioning. A typical paper in, say, semantics of conditional statements in natural language might start with a description of intuitively valid and invalid reasoning forms, and then propose some underlying meaning for the key construction "if then" which shows how the valid inferences come about, while avoiding the invalid ones. Here is a concrete example. Consequents may be weakened freely: "if A, then B" implies "if A then B-or-C" – but antecedents cannot be strengthened freely: "if A, then B" does not always imply "if A-and-C, then B." (Let A be "I put sugar in my coffee," B "the coffee tastes fine," C "I put Diesel oil in my coffee"). Standard propositional logic does not explain this fact, as the truth tables will validate both inferences.

But the literature contains various proposals drawing a natural distinction. For example one often postulates a preference relation among situations (say of "relative plausibility"), and then calls a conditional "if A, then B" true if B holds, not in all situations where A holds (which yields classical entailment), but only in all *most-preferred* situations where A holds. Other accounts of conditionals exist, too, providing "dynamic" alternative views of the same phenomena (including one by Grice). This is how far logical analysis usually goes. One systematizes the observable facts of language use by postulating certain underlying mechanisms, which may involve "theoretical terms" that have no direct surface reflection. (Conditionals do not refer to preferences explicitly.) The latter abstractions may then be used in other contexts too: e.g. preferences of various kinds are gaining popularity in semantics.

Logical theories of this sort can have great sophistication, and broad sweep. But what they do not attempt to explain is how their key structures came about, or why they function the way they do. These more ambitious questions are generally considered hard, having to do with underlying mechanisms of cognition that we do not yet understand. Many people (including myself) believe that natural language strikes some kind of successful balance between *expressive power* for information content and *resource complexity* of cogitation and communication processes. But what do we mean by these terms in the absence (so far) of a truly fundamental theory of information and cognition? And another doubt: perhaps searching for deep explanations may even be inappropriate. After all, human languages are the result of a long evolution, with many historical accidents, where things could easily have gone another way. We cannot "explain" human history. Why should we be able to "explain" language?

As opposed to this, Rubinstein offers a staunchly *a prioristic* analysis of various linguistic phenomena, trying to show how things function the way they do by being kept firmly in place through the Invisible Hand of optimization and strategic equilibrium. Economists in this mode are Leibnizian optimists: we live in "the best of all possible worlds," or failing uniqueness, at least "in an optimal possible world." I myself love this kind of analysis (no matter its eventual prospects of success), which is in the same spirit as shortest-path principles in physics, or other grand ideas driving science. Logic can profit from adding such concerns!

What does one want to explain about language? Let's be ambitious: (a) its structure (why it contains the expressions that it does), (b) the steady state of its competent use (the rules that guide communication), (c) the mechanisms for learning these skills (which take time), (d) the robustness of linguistic practices, but also (e) the possibilities of failure, and then revision. The book offers a number of examples. One chapter explains certain lexical vocabulary, one analyzes speech acts (warnings, and reactions to them), a third rules of dialogue (procedures for argumentation). The types of explanation are different in each case, but I'll come to that. Of course, at this level of abstraction, we do not expect very precise explanations for specific linguistic facts. Indeed, the book contains hardly any of the standard household items for linguists or logicians. Nevertheless, what is offered roughly follows the standard division into syntax, semantics, and pragmatics. And since there are no miracles, the simplicity of the game-theoretic analysis will necessarily be matched by the generality of the conclusions.

3 Explaining code

I will give my understanding of what Rubinstein does in his case studies, and then compare with some logical issues. The first batch of examples in chapter 1 of the book concerns linguistic code – namely, lexical meanings expressed by ordering predicates. Here analogies with standard logical notions and approaches are clear, and the links are obvious.

Why are certain orderings so ubiquitous?

According to the author, certain types of ordering relation are especially prevalent in natural language – transitive linear orderings and more generally, complete asymmetric orderings or tournaments. He explains the first phenomenon as follows. (Arguments always take place in some fixed finite model.) Linear orders allow one to describe each individual in a domain uniquely via first-order logic, in terms of its position in the ordering. He then converts this observation (a typical mentality of a mathematician!), and shows that *only* linear orders have this unique identification property. In other words, given the importance of unique reference, language must provide for it, and indeed it does. With tournaments, Rubinstein's analysis is different, not in terms of maximal expressiveness, but minimal complexity. He considers definitions for binary relations R that we want to communicate by means of language to others. Clearly, minimal complexity is a bonus here. One natural description format is this. Describe some general properties of R (again, in first-order formulas, such as those defining a tournament) and give some concrete examples of ordered pairs that fall under it. Complexity can then be measured as the minimal number of concrete instances one needs on top of some first-order description. He shows (*modulo* some technicalities) that the relations of lowest complexity in this sense are the tournaments – and, again, we are not surprised to see that language employs these in abundance. There is also fascinating stuff about approximating relations in terms of hits-and-misses, but my aim is not to abstract the whole book.

Explanations for vocabulary

These analyses are very much in line with standard logical analysis, even though I have never seen quite these results in books on model theory. For example, consider the tournament example. Basically, a general first-order description of a binary relation can only describe this up to its "isomorphism class": all binary relations on the given model sharing its first-order properties. To identify further, one needs to give some specific instances. Logicians have rather focused on the size of such isomorphism classes: the other side of the same coin, given that we are talking about

Comments

(with n objects in our model) $2n^2$ binary relations altogether. As a minor quibble, I would certainly also want to count the size of the general first-order formula as a parameter of the complexity, not just the number of instances. If we do that, we also get close to general principles in logic and computer science of so-called *minimal description length*. This is often implemented in terms of the "Kolmogorov complexity" of an object – i.e. the minimal size of a program for some Turing machine that outputs the object, in some effective representation. I think that Rubinstein's analysis fits most naturally into this general framework. Also, I think one can say more about the general first-order part. Not all first-order descriptions are equal. One most effective device for saving on concrete instances are "Horn clauses": universal implications from conjunctions of atoms to single atoms, such as transitivity: $\forall xyz\ ((Rxy$ and $Ryz) \rightarrow Rxz)$. These automatically produce further instances deterministically. Significantly, Horn clauses are the computational part of first-order logic, exploited in logic programming.

But logic has other methods than these for explaining the occurrence of vocabulary – even though they are seldom presented together under that heading. I shall mention a few basic ones.

Type 1: *natural invariances and their linguistic reflection*
Nature around us shows certain invariances – e.g. under Euclidean transformations of space (translations, rotations). It is an old idea of Helmholtz in the nineteenth century that such invariances will be naturally reflected in language, which reserves basic lexical expressions for operations or predicates that are invariant under these transformations (such as geometric "betweenness" – incidentally, a basic ternary, not binary relation). There is a whole bunch of results in logic showing how logically definable predicates in various languages coincide with invariant ones of different kinds. For example, the basic logical expressions (Boolean set operations) are invariant for arbitrary permutations of individuals in a domain. Further categories of expression in natural language, such as

spatial prepositions ("inside," "behind," "toward") will be invariant for other classes of more geometrical transformations. This type of invariance analysis occurs in the present book, too – but in connection with schemata of definition for social choice.

Type 2: *lowest complexity and its expressive reflection* Especially in the logical literature on so-called generalized quantifier expressions ("some," "all," "ten," "most," "few," "enough"), there is a body of results of the following kind. Expressions of quantification may be associated with computation procedures that determine, given the objects in a domain, whether the numerical truth condition of the quantifier is satisfied. These procedures can be associated with suitable automata. First-order quantifiers can be computed with finite-state automata, others, like "most," rather require push-down store automata with unlimited memory. Now, it can be shown that natural language provides quantifier expressions for all "semantic automata" of lowest complexity. Of course, quantifiers are just one type of expression, close to logic. It would be much harder to explain the vocabulary in other categories of language, such as ordinary verbs (mathematicians once attempted a classification of verbs via Catastrophe Theory) – but at some stage, we must also leave room for accidents of our environment!

Note the two broad lines of thinking here. The first explains vocabulary as a reflection inside our cognitive apparatus of invariances occurring outside of us in nature. The second explains vocabulary as a result of computational limitations inside our cognitive apparatus. Both seem attractive, and there need not be one uniform source. But there are still further approaches! The preceding was about individual agents, while we might also explain vocabulary through the necessities of social interaction (more congenial to game theory). Perhaps the oldest result of this type is the *Dutch Book Theorem* from the 1950s, which explains the usual meanings of the propositional connectives in probabilistic reasoning as the only ones that do not allow the making of unfair "Dutch books" against one's opponents. This result has remained isolated, though (although Ariel Rubinstein informs me that there has been some game-theoretic follow-up) – but I have often wondered about similar analyses for other logical notions.

Discussion

My conclusion is that Rubinstein's analysis in chapter 1 is very close to logic indeed – though certainly raising nice new questions. But as we have seen, there are many approaches in this vein, and one should probably discuss their comparative merits. On the critical side, I just mention one empirical point. Are the author's observations correct? Are linear orders or tournaments indeed the basic stock of natural language? I doubt it. To me, the most obvious linguistic category of binary relation are *comparatives*. These are so basic that language even has a systematic operation for building them: from "large" to "larg-er." But the complete formal properties of comparatives are known: they are asymmetric and "pseudo-connected": $\forall xyz \ (Rxy \rightarrow (Rxz \lor Rzy))$. This form of completeness is neither of Rubinstein's two cases. So, I would say he still owes us an equally clever account of comparatives.

4 Explaining speech acts

The next example (chapter 2), much more game-theoretic, is an account of the emergence of warning signals, uttered by speakers, and interpreted and then acted on appropriately by their hearers. This is closer to linguistic pragmatics, although it also involves semantics. The pattern of explanation here comes from evolutionary game theory, illustrated by a practical example. Basically, we can explain the existence of warning speech acts as follows. If the language did not have such a facility, a group of "mutants" could improve their communicative and practical situation by endowing some expression with this meaning, and utilizing the warnings amongst themselves. This phenomenon would then presumably spread eventually to all users. Nevertheless, there is a difficulty. One cannot simply tamper with an existing language. If the addition causes nonmutants to respond in harmful ways (e.g. panic as a response to the new warning signal), then the mutants may be worse off on average. Rubinstein then explains in progressively more technical terms how such harmful reactions will not occur, provided we assume that preferences of language users do not just count utilities, but also the *complexity* of following a strategy. In particular, panic is costly, so nonmutants will not bother.

I cannot pretend I got all the ins and outs of this, but this general pattern of explanation is very interesting. Language eventually reaches some kind of maximal expressiveness, as measured by useful communicative functions. Nevertheless, there are many arbitrary assumptions in the example, and a more real challenge for me would be to provide accounts involving similar assumptions for more central features of communication. In particular, although Rubinstein discusses Grice's Maxims of informativeness and relevance, he refrains from providing a game-theoretic underpinning for them. Now *that* would be something! Another thing is this. We can distinguish between the performance of language users in "steady state" of mature competence versus their *learning* processes. How does the game-theoretic account relate to this crucial distinction in understanding how language functions? Finally, I would like to see some relation between the pattern of explanation given here and the more model-theoretic style of analysis discussed in chapter 3. Take the issue of expressive complexity, which seems related to computational complexity for language users in getting the meanings right. I think complexity is over-riding: and language must do with what is feasible, rather than what is desirable. Thus, in many cases, one would like to invert Rubinstein's lexicographic preference for utility over complexity.

Finally, even though there are no direct analogues to this evolutionary game analysis in logic, there are congenial themes. For example, David Lewis' famous account of the emergence of linguistic and nonlinguistic *conventions* has a similar slant: what coordination moves will groups of agents develop to achieve certain goals? In the subsequent literature, much emphasis has been placed on what the agents need to *know* about each other in order to communicate effectively. I sense similar epistemic aspects to the mutant thought experiment (e.g. mutants might come to recognize mutants via signals, so as to exclude harmful confrontations with "old-timers"). But these remain implicit in the analysis presented in the book.

5 Explaining rules

A third major example in the book are argumentation conventions (chapter 3). Rubinstein observes how argumentation has a game-like structure, with moves and procedural conventions. He then turns to specific empirical phenomena in "truth-finding." In particular, one challenge is to explain why certain arguments or counter-arguments seem more persuasive than others to certain claims, even when there is no obvious preference in terms of informative value. Suppose someone claims that in most capitals of the world, education levels are rising. Your opponent challenges you and points out that it has sunk in Bangkok quite recently. You have positive evidence concerning Manila, Cairo, and Brussels, but can present only one of these. Most people agree that you should mention Manila. But why? Rubinstein gives an intriguing analysis of debates and persuasion rules which shows that optimal debates between players presenting conflicting evidence to a "listener" are sequential, and have a persuasion rule that must treat some arguments i, j symmetrically: i is persuasive against j, but also vice versa.

This example is again very significant from a logical point of view. Argumentation and dialogue games have existed in logic since the 1950s. Roughly speaking, valid conclusions are those assertions C for which their proponent has *a winning strategy in any debate* against an opponent granting the premises. In recent years, these games have been taken up by computer scientists who use their procedural, as well as compositional structure for modeling interactive processes. This all concerns logical validity, and these games are very regimented. Little work has been done on games with more relaxed rules that would model common sense reasoning or heuristics, let alone informal argumentation. Rubinstein's new style of analysis may point in just the latter direction. Even so, I had some trouble locating a crisp conclusion concerning the cities, because the author's moral seems to depend on the debate rules in the end. But this may be just as well, because the observed phenomenon is incomprehensible from a purely informational point of view – so there must be some additional rationale based on implicit procedural conventions.

6 How does natural language function?

The three examples display very different explanatory strategies. Indeed, they are very concrete, and specific assumptions about possible moves and utilities are cooked up ad hoc. This book does not attempt to provide a *systematic* view of natural language. I do not find this an objection, given its originality. But one would like to see if accounts like this would work more generally, for other orderings, speech acts, or argumentation conventions. Also relevant is this. At least some empirical facts about language use and complexity are known. Zipf's Law, for example, tells us that the highest-frequency expressions tend to be the lowest-complexity ones – as one would expect of an optimal medium. This should surely be relevant to game-theoretic analysis. Also, linguists and logicians have their own favorite Grand Challenges concerning natural language, clamoring for new breakthrough ideas, and it would be of interest to see if this kind of game theory has something to offer on those. For example, what is the game-theoretic explanation for the ubiquity of the phenomenon of *ambiguity* in meanings?

I have two specific recommendations for broader themes. One concerns a convergence of interests. As I said before, *preferences* are an obvious theme in the intersection between game theory and modern logic and linguistics. One currently influential program in linguistics, for example, is Optimality Theory, which attempts to explain syntax and semantics through the selection of packages of constraints that are most-preferred in some suitable sense. (At lower unconscious levels of sentence comprehension, these preferences will be hard-wired – whereas they are consciously manipulable at the higher level of discourse.) These analogies call for interaction. Another recommendation is this. Since Rubinstein's game-theoretic analysis often concerns "what if" situations where people change a language, it may be important to analyze actual processes of *new language formation*, which occur around us all the time – most notably, in the design of specialized natural, mathematical–logical, or computer programming languages.

7 Terms and formats in game theory

Part 2 of the book contains some analyses of the role of language in game theory. With some of this, I am not even sure that "language" is the right heading, because it seems to concern general methodological issues, often similar to what philosophers of science have long worried about. For instance, the status of terms in scientific theories has been debated since the Vienna Circle. "Observational terms" are correlated with empirical observations, while "theoretical terms" serve the purposes of abstract organization, but do not correspond directly to observation or common sense. A good example are terms like "mass" or "energy" in physics, which bear only an oblique resemblance to their original common-sense meanings. Very briefly, Rubinstein's worries about the theoretical term "strategy" seem to dissolve then – the term is not the same as the common-sense notion, but is quite legitimate all the same. What remains is the warning that rhetorical uses of game theory overlook these subtleties.

A more concrete and technical subject is the discussion of *definable preferences*. Rubinstein points to the importance of linguistic formats for natural definitions of individual and collective preferences. He takes these from various first-order formalisms, rather than the quantitative mathematical formulas often employed by game theorists, which may lack reflection in any plausible description language. Results concern the use of constraints narrowing down the admissible forms of definition, such as desirable ordering properties, or suitable invariances between 'dating profiles' of candidates for a person seeking a partner. I will content myself with saying that chapter 4 is straight logical model theory, which could also have been published in a logic journal! Admissible preferences are defined in terms of their invariance behavior (e.g. with respect to permuting individuals up to sameness of profiles), with an interesting use of the interpolation theorem to ensure uniformity of definitions across domains of different sizes. What I would like to advocate here is joining forces. Social choice theory, game theory, logic and linguistics have obvious shared concerns here. Let me mention two current ones from my world. What is the dynamics of *changing preferences*? After all, informational preferences may change. I may consider all four possibilities for two propositions A, B equally likely, without any preference at the start. But then, upon hearing a default conditional statement "if A, then B," I upgrade this preference, making the A/not-B case less preferred than the others. In the end, we want dynamics, not just statics of preferences. Another point is the general move in semantics from one-agent to *many-agent* settings – e.g. moving from individual knowledge to common knowledge. This move is clear in game theory, where collectives can form coalitions (cf. the "mutants" in an earlier-mentioned story). All these concerns seem to fit into one coherent picture: but which?

8 Game theory and logic

I guess my conclusion is abundantly clear. I find this book very congenial, overlapping with logic all the way through, but very original precisely where it diverges. Of course, I would like to see more systematics in methods and aims, plus more grappling with the broader issues that drive logic and the semantics of natural language. But that seems feasible, and promising. Indeed, one can apply the same game-theoretic style of analysis to many topics

Comments

inside logic proper, it seems to me. (Why do we use the proof principles that we do?) As for the soul-searching aspect, I also see similarities. Logic is about reasoning, but it never quite fits the observed phenomena. If the latter diverge too much, logicians happily change their tune, and say that logic serves to *improve*, rather than describe practice. And indeed, in cognitive science, there is this very real phenomenon that theory can influence behavior – which makes the discipline much more subtle than the natural sciences. So I'd say: "game theorists, relax!"

REFERENCES

van Benthem, J. (1986) *Essays in Logical Semantics*, Dordrecht: Reidel

(1996) *Exploring Logical Dynamics*, Stanford: CSLI Publications

(1998) "Logical Constants: The Variable Fortunes of an Elusive Notion", in R. Sommer and C. Talcott (eds.), *Proceedings Feferfest*, Stanford, forthcoming

(1999) *Logic and Games*, electronic lecture notes, University of Amsterdam, *http://www.turing.uva.nl/~johan/teaching*

van Benthem, J. and A. ter Meulen (eds.) (1997) *Handbook of Logic and Language*, Amsterdam: Elsevier

Doets, H.C. (1998) *Basic Model Theory*, Stanford: CSLI Publications

Gärdenfors, P. (1988) *Knowledge in Flux*, Cambridge, MA: MIT Press

Gärdenfors, P. and H. Rott (1995) "Belief Revision," in D. Gabbay, C. Hogger and J. Robinson (eds.), *Handbook of Logic in Artificial Intelligence and Logic Programming, IV*, Oxford: Oxford University Press, 35–132

Gilbers, D. and H. de Hoop (1998) "Conflicting Constraints: An Introduction to Optimality Theory" *Lingua*, 104, 1–12

Kemeny, J. (1955) "Fair Bets and Inductive Probabilities," *Journal of Symbolic Logic*, 20, 263–73

Kuipers, Th. (1999) *Structures in Science. Heuristic Patterns Based on Cognitive Procedures*, Philosophical Institute, University of Groningen, Synthese Library, Dordrecht: Kluwer, forthcoming

Lewis, D. (1969) *Convention*, Cambridge, MA: Harvard University Press

Johan van Benthem

Li, M. and P. Vitányi (1993) *An Introduction to Kolmogorov Complexity and its Applications*, New York: Springer-Verlag

Prince, A. and P. Smolensky (1997) "Optimality Theory: From Neural Networks to Universal Grammar," *Science*, 275, 1604–10

Shoham, Y. (1986) *Reasoning about Change*, Cambridge, M.A.: MIT Press

Veltman, F. (1996) "Defaults in Update Semantics," *Journal of Philosophical Logic*, 25, 221–61

ECONOMICS AND LANGUAGE

Tilman Börgers

The quality of this book confirms that Ariel Rubinstein is one of the most important game theorists, and, more generally, theoretical economists today. His importance lies in his originality. He urges researchers to break out of established thought patterns. He opens up new lines of thought. He is willing to disagree with predominant views. All these qualities are reflected in the current book.

As their author readily acknowledges, some chapters of this book are loosely, but not tightly connected. Chapters 1, 2 and 3, however, have much in common. They are all based on the claim that the ideas of game theory can be fruitfully applied to the study of language. Reading and reflecting on this book has convinced me that this is indeed the case. The best evidence which I can offer is the fact that while reading the book I could think of many additional ideas in this area which appeared to me as promising research subjects. Some of these ideas will be indicated below. Given the limitations of space for this comment, and because of the coherence of chapters 1–3 I shall focus in my discussion on these three chapters.

The simplest game-theoretic model of language is described in chapter 2. A parable can capture the essence of this model: Exactly one of several boxes contains a prize which individuals A and B can share if they find it. Individual A, but not individual B, observes the box into which the prize is placed. However, it is individual B who has to open a box, and B has only one chance. Before B chooses A can make one of several sounds, which can be heard and distinguished by A.

I am grateful to V. Bhaskar for helpful discussions.

One equilibrium of this game is that a "language" develops: There is a one-to-one relation between the sounds made by A and the boxes, and A uses the language to truthfully indicate which box contains the prize. B then opens that box. Unfortunately, there are other equilibria of this game. For example, A may make a random sound, which B then ignores. This is an equilibrium because A has no reason not to make a random sound if he is ignored by B, and B has no reason to pay attention to A, given that A's utterance is random. This equilibrium is, of course, less attractive to A and B, but that doesn't mean that it won't be played. Another bad equilibrium is that A *always* says "The prize is in box *b*," independently of where it really is, and that B, when hearing this claim, opens some arbitrarily selected box, and otherwise, if A does not make this claim, refuses to open any box at all.

If we wish to provide a game-theoretic account of language, we need an explanation of why a communication equilibrium rather than one of the bad equilibria is played. Chapter 2 describes an argument which enriches the model by introducing complexity costs, and which refines the equilibrium notion involving evolutionary considerations. A key argument is that an evolutionary process will move out of the bad equilibrium because a random mutation will appear where some individuals use and interpret signals accidentally in the same way. Given the enormous number of sounds which humans can make, this looks like an extremely unlikely coincidence. Chapter 2 leaves me unconvinced that the multiplicity problem in language games has been resolved.

A variation of the simple model of chapter 2 has been considered in the "cheap talk" literature: A and B both talk in stage 1, and take actions in stage 2. However, A and B have no factual information to communicate. They just talk. The second-stage game is a coordination game with, say, two equilibria, one of which is better for both A and B than the other. Cheap talk games have equilibria in which anything that is said in the first stage is ignored in the second stage. But they also have equilibria in which both players say X in stage 1, and play the good equilibrium in stage 2 if and only if both agents said X in stage 1. Evolutionary arguments[1] have been proposed to select equilibria of the latter type (Kim and Sobel, 1995).

I mention cheap talk models because they contrast in an interesting way with the model in chapter 2. Whereas in chapter 2 speech serves the purpose of communicating observed facts, in cheap talk games speech serves the purpose of reinforcing some particular action. Cheap talk consists thus of ethical (or other) imperatives. Cheap talk models might provide the beginning of an evolutionary rationale for the large amount of ethical talk in which we all engage every day.

Chapter 1 is concerned with the more ambitious task of explaining the structure of language in detail, again using game-theoretic models. I would like to view the model in chapter 1 as an extension of that in chapter 2, although this is not how Rubinstein presents chapter 1. Imagine that in our earlier parable the set of boxes in which the prize may be can change. Sometimes there are boxes b, b' and b'', etc. At other times, there are only boxes b' and b''. Suppose that before the communication game begins individuals A and B can see which boxes are there, and that they also can see how these boxes are arranged. Specifically, assume that for any two boxes it is observable whether box b is to the left of box b' or vice versa. Finally, assume that another observable relation is whether all colors which appear on box b' also appear on box b or vice versa.

My interpretation of chapter 1 is that it is concerned with an analysis of equilibria of this game, although this game is not explicit in chapter 2. As in chapter 1, the set of equilibria of the communication game implicit in chapter 2 is large. Rubinstein implicitly focuses on those equilibria of this game in which words have meaning and communication takes place. He also invokes a further exogenous restriction regarding the equilibrium language. The language must identify objects only by use of a single binary relation. The question on which Rubinstein focuses is which of the available binary relations will be used.

---

[1] Unlike chapter 2, these arguments do not involve complexity costs.

The answer which Rubinstein proposes is that it will be linear orderings, such as the binary relation "to the left of." An incomplete ordering, such as "all colors which appear on b' also appear on b," by contrast, will not be used. Rubinstein puts forward three reasons for this. One reason is that only linear orderings allow unique identification of objects in any set of possible objects. Another argument is that linear orders are particularly easy to teach, and hence easy to transmit. Another argument fits less well into my context: Linear orders are well suited to describe other binary relations.

I want to emphasize two features of the argument in chapter 1. First, to understand the structure of language a game model is constructed in which some information is commonly observed (in my parable: the set of boxes in which the prize might be, and their relations) and some information is privately observed (the particular box which contains a prize). The language can use phrases (such as "the leftmost box") which derive their specific meaning from the commonly observed context. Almost all phrases of our languages have this property that they don't refer to some specific individual, but can be understood only when some commonly observed context provides background information.

The second feature of chapter 1 that seems worth noting is that implicitly an equilibrium selection argument is used. This argument invokes two ideas. One is that language develops so that it is functional – i.e. useful to society. The second is that language develops so that it is easy to transmit. The first half of this is even announced as an integral part of the "research agenda": "features of natural language are consistent with the optimization of certain 'reasonable' target functions." As indicated earlier, I have reservations about this argument. Ease of transmission, though, seems to me a very natural criterion to invoke when selecting among equilibria, although it might easily be in conflict with functionality.

Arguments similar to those in chapter 1 can be used to investigate other, possibly more basic issues. Suppose no binary relations were observed. The equilibrium language could still make use of the fact that the set of boxes is commonly observed. For example, it might be that, although many of the boxes are made of cardboard, typically only one of the boxes placed in front of B is of cardboard, and all others are of wood. Then a functional and easy to transmit language might develop a single word to refer to cardboard boxes without differentiating between individual boxes. Thus one might start to construct a theory which explains why language so often uses category names to refer to individuals.

Chapter 3 could also be fitted into a version of the framework that I have used to understand chapters 1 and 2. Unlike in chapters 1 and 2, however, in chapter 3 both individuals have some private information. Moreover, now their interests differ. The question which Rubinstein asks is how a debate between A and B should be structured if it needs to be short, and if we wish to elicit as much information as possible.

The step from chapters 2 and 1 to chapter 3 is quite large. One might make a less ambitious step, and stay with a model in which A's and B's preferences are identical. Consider, for example, the model of chapter 2, and suppose we enriched this model by just one ingredient, namely the assumption that A and B both have pieces of private information. Suppose it were exogenously given that A spoke first, then B spoke, and then A spoke again.

One equilibrium of this game would involve A saying everything that he knows, followed by two rounds of silence in which neither agent says anything. This would be efficient, in the sense that B could make a choice based on all available information. But now suppose that speaking, and listening, is not without costs. Then both agents might be better off in another equilibrium in which A lets B speak first, and then establishes which of his own private observations are actually relevant, in the sense that they could improve B's choice, given what B already knows.

An analysis of equilibria of common interest games with initial sequential talk might lead to the beginning of a theory of *conversation*. Conversation surely differs from *debate* in that the participants pursue common rather than opposed interests. Grice's *cooperative* principles, quoted in Rubinstein's chapter 3, seem more apt for conversation than for debate. The above argument might provide a formal counterpart to Grice's third maxim ("Relation: Be Relevant"). I find Grice's four maxims of quantity, quality, relation, and manner highly appealing, and have secretly been toying with the idea of posting them in seminar and conference rooms. Among other things, they suggest, though, that I should conclude my contribution here.

REFERENCE

Kim, Y.G. and J. Sobel (1995) "An Evolutionary Approach to Pre-play Communication," *Econometrica*, 63 1181–93

ECONOMICS AND LANGUAGE

Barton L. Lipman

"Is he – is he a tall man?"
"Who shall answer that question?" cried Emma.
"My father would say, 'Yes'; Mr. Knightly, 'No';
and Miss Bates and I, that he is just the
happy medium."
(JANE AUSTEN, *EMMA*.)

1 Introduction

While a reader of this book may be surprised to see a game theorist writing about language, he should instead be surprised by how few game theorists have done so. As Rubinstein observes, language is a game: I make a statement because I believe you will interpret it in a particular way. You interpret my statement based on your beliefs about my intentions. Hence speaker and listener are engaged in a game which determines the meaning of the statement.

Furthermore, I think more "traditional" economists should be interested in models of language. The world people live in is a world of words, not functions, and many real phenomena might be more easily analyzed if we take this into account. For example, consider incomplete contracts. Our models treat contracts as mathematical functions and hence find it difficult to explain why agents might not fully specify the function. Of course, real contracts are written in a language and may not unambiguously define such a function – not its domain or range, much less the function itself.

I thank Marie-Odile Yanelle for helpful conversations. No thanks are due to the Dell Corporation.

Rubinstein gives an intriguing opening to this important topic. In what follows, I give a brief overview of what I see as the important ideas in the book and use this as background to comments on directions for future research. In the process, I will suggest that there are important and difficult problems ahead.

2 Overview

While Rubinstein jokes that there is little to connect his lectures beyond the common title and speaker, I think it's conceptually useful to consider the various ingredients of a theory of language the book gives. As I see it, chapters 1–4 give us various ways to address three questions:[1]

1 What gives a statement its meaning?
2 Why is language the way it is?
3 How does language affect actions?

Rubinstein gives two distinct ways to answer the first question. Chapters 1 and 4 give one of his approaches, modeling language as a *logical system*. In this approach, language is a set of building blocks and rules of construction which implicitly define the meaning of statements. For example, the meaning of $p \lor q$ is that either p or q is

---

[1] I have only one comment on chapter 5. Rubinstein criticizes the way game theory is "sold" to business students, saying that the kind of conceptual understanding that game theory can provide "stop[s] far short of giving advice and claiming to be useful." I agree up to a point. Having taught game theory to MBA students for a few years, I spent much time reflecting on what value, if any, the theory had for them. I think most business students want rules, not concepts, evidently believing that the business world is a series of clearly framed problems, each with an appropriate rule to follow. This absurd notion influences the way textbooks are written since these are supposed to appeal to business students. I disagree with Rubinstein, though, when he concludes that there is nothing "useful" to such people in game theory. I think it is precisely the concepts and understanding he alludes to which can benefit someone in business. Of course, their understanding will never approach Rubinstein's! In particular, backward induction is indeed a revelation to business students and the concept is hard for them to learn. Any reader who is skeptical on this point need only see how many business students will err on a question involving sunk costs.

true – a meaning which, of course, depends on the underlying meaning of p and q.

While this is a natural starting point, it is clear that language as spoken by real people is *not* logical in the formal sense. For example, as Rubinstein notes in chapter 3, a listener generally makes inferences in response to a statement which go beyond the purely logical implications of what the speaker said. Put differently, the fact that a given statement is made can itself signal some of the speaker's private information. Later, I will discuss another way in which language is not purely logical.

Rubinstein's second way to answer question 1 addresses this drawback of the logical system approach, adopting instead what I'll call the *equilibrium approach*. This approach, used in chapters 2 and 3, treats language as a set of words and an equilibrium interpretation of their usage or meaning. For example, if a particular statement is made only in certain situations, then the meaning of the statement (in addition to any concrete evidence included, as studied in chapter 3) is that one of these situations must be true. While the equilibrium approach captures some of what the logical system approach misses, it does so at a cost: the language has no structure to it. Hence many questions about language cannot be addressed in such a model.

Rubinstein's approach to the second question, illustrated nicely in chapters 1 and 3, is what I'll call the *structural optimization hypothesis*. In this approach, language is that structure which maximizes the amount of information conveyed subject to constraints on the "complexity" of the language. In chapter 1, the "complexity constraint" is that the language can have only one relation. The question then is what relations are most useful for describing many objects (or sets of objects) as precisely as possible. The complexity constraint in chapter 3 is that the listener cannot process more than two pieces of evidence. Here Rubinstein considers the optimal way for the listener to interpret statements by debaters in order to elicit information from them.

Finally, chapter 4 gives an intriguing approach to answering the third question which I'll refer to as the *expressibility effect.* The general perspective here is that people perceive the world and form decisions in words. Because of this, the nature of the language people use affects their actions. Chapter 4 looks at the expressibility effect through the requirement that decision rules or preferences be definable but, as Rubinstein notes, the principle could be explored in many other ways and other aspects of decision making.

3 Further research

I now turn to three areas for future research in this area which strike me as particularly interesting. First, I am intrigued by the inherent circularity one gets when combining the structural optimization hypothesis and the expressibility effect. By means of illustration, one response to chapter 1 which I have heard is that it may be that linear orderings are common in our language not because of any inherent usefulness they have but simply because linear orderings are common in the natural world. I think this criticism misses a crucial point: do we perceive linear orderings to be common in the world because they are common in our language? In other words, the structural optimization approach suggests that we structure language in a way which seems useful to us given our perception of the world. On the other hand, because of the expressibility effect, once developed, language affects the way we see the world. An analysis which includes such feedback effects could be quite interesting.

Second, as noted earlier, both the logical systems and equilibrium approaches to meaning miss part of the picture. It is clear that language does have structure even if it is not fully logical. It is not obvious how to model this simple notion. If the meaning of a sentence comes from its equilibrium interpretation only, then there is no reason for structure to relate to content. "I live in Wisconsin and it is cold" could be interpreted as the conjunction of "I live in Wisconsin" and "It is cold" or the disjunction of "I am hungry" and "Rubinstein lives in Israel." At the same time, meaning cannot come purely from the content of the sentence in isolation from its context or other extra-logical factors. A model which combines the advantages of equilibrium models and logical systems would be more plausible – and, perhaps, more useful.

It might be possible to develop such a model using the structural optimization approach. In this case, a particular natural complexity constraint is limited memory. It seems intuitively obvious that it is easier to remember that "I live in Wisconsin and it is cold" means "I live in Wisconsin" and "It is cold" instead of some notion unrelated to Wisconsin or cold. Similarly, the structure of words might be derivable from memory limitations. For example, consider the use of prefixes and suffixes. To deduce the meaning of a word beginning with "post," it is sufficient to remember the meaning of the prefix "post" and the word which follows. To combine this with an equilibrium approach to meaning, one would presumably focus on "efficient" equilibria.

The last topic is another aspect of the issue of meaning: vagueness.[2] Perhaps it is easier to understand what I mean by "vague" by contrasting it with "precise." I will say a term is *precise* if it describes a well defined set of objects. This is the way language works in most of the economic models where it appears: the set of objects is partitioned and a word is associated with each event of the partition. By contrast, a term is *vague* if it does not identify such a set.

To illustrate my meaning and to show why vague terms make it difficult to model language as a logical system, consider the following version of the famous *sorites paradox*. Two facts seem clear regarding the way most people use the word "tall." First, anyone whose height is 10 feet is tall. Second, if one person is tall and a second person's height is within 1/1000 of an inch of the first, then the second person is tall as well. But then working backward from 10 feet, we eventually reach the absurd conclusion that a person whose height is 1 inch is tall. Of course,

---

[2] Keefe and Smith (1996) is a fascinating introduction to the philosophy literature on the subject.

the source of the difficulty is clear: "tall" does not correspond to a clearly defined set. There is no fixed height which defines the line between someone who is tall and someone who is not. Because of this, there is an inherent ambiguity in the meaning of the word "tall" as the quote I began with illustrates. Many other words have this property: consider "bald," "red," "thin," "child," "many," and "probably."

The prevalence of vague terms in natural language poses two intriguing and inter-related challenges. First, what meaning do such terms convey? Second, why are they so prevalent?

As to the first question, it seems clear that vague terms acquire their meaning from usage. That is, whatever meaning "tall" has is due to the way people use the word, not any particular logical structure. Hence it seems natural to follow the equilibrium approach to studying meaning. The most obvious way to do so is to treat vague terms as ones which are used probabilistically, so the probability someone is described as "tall" is increasing in height but is strictly between 0 and 1 for a certain range. This kind of uncertainty could correspond to mixed strategies or private information, either of which would give a clear notion of the meaning of a vague term.

However, this approach has a severe drawback: it cannot give a good answer to the second question. In particular, if this is what vagueness is, we would be better off with a language which replaced vague terms with precise ones. To see the point, consider the following simple example. Player 2 must pick up Mr. X at the airport but has never met him. Player 1, who knows Mr. X, can describe him to player 2. Suppose that the only variable which distinguishes people is height and that this is independently distributed across people uniformly on $[0,1]$. 1 knows the exact height of Mr. X; player 2 does not. However, both know that when player 2 gets to the airport, there will be three people there, Mr. X and two (randomly chosen) others. Player 2 has very little time, so he can ask only one person if he is Mr. X. If player 2 chooses correctly, 1 and 2 both get a payoff of 1; otherwise, they both get 0. Clearly, there is a simple solution if 2 can observe the heights of the people at the airport and the set of possible descriptions is $[0,1]$: 1 can tell 2 Mr. X's height. Because the probability that two people are the same height is zero, this guarantees that Mr. X will be picked up.

However, this is a much bigger language than any in use. So let us take the opposite extreme: the only descriptions 1 can give are "short" and "tall." Further, player 2 cannot observe the exact height of any of the people at the airport, only relative heights. That is, player 2 can tell who is tallest, who is shortest, and who is in the middle.

In this case, it is not hard to show what the efficient language is: 1 should say "tall" if Mr. X's height is greater than $1/2$ and "short" otherwise.[3] Player 2 tries the tallest person at the airport when he is told that Mr. X is "tall" and the shortest when told that Mr. X is "short." Note, in particular, that there is no vagueness in the optimal language: "tall" corresponds to the set $[1/2, 1]$.

What would "vagueness" mean in this context? One way to make "tall" vague would be if 1 randomizes. For example, suppose 1 says "short" if Mr. X's height is below $1/3$, "tall" if it is greater than $2/3$, and randomizes in between with the probability he says "tall" increasing with Mr. X's height. While this resembles the way "tall" is used in reality, there are no equilibria of this kind. If one takes a nonequilibrium approach and assumes player 1 is committed to such a language, it is easy to show that both player 1 and player 2 (and presumably Mr. X!) would be better off in the pure strategy equilibrium above.[4]

Alternatively, private information could give the needed randomness. Suppose player 1 has observed some signal in addition to height. In this case, the efficient language partitions not the set of heights but the set of height–signal pairs. Hence a word will correspond to a random statement about height, the randomness being induced by the signal.

---

[3] Of course, the words themselves are not relevant to this equilibrium. An equally efficient language would reverse the roles of "tall" and "short," or even replace them with "middle-aged" and "blond."

[4] Lipman (1999) gives a generalization of this argument and some further discussion.

On the other hand, what is this other signal? First, suppose it is somehow intrinsically relevant. For example, player 1 may know Mr. X's weight and player 2 may be able to observe relative weights. In this case, player 1 needs to communicate on two dimensions to player 2. The efficient language will have terms which are precise in two dimensions even though this may make them imprecise in any one dimension. This does not seem to be much of an explanation of an apparently unidimensional term like "tall." On the other hand, suppose the signal is not relevant. Then it would be most efficient to ignore this signal and use the language described above.

Why, then, are vague terms so prevalent? It seems rather obvious that such words are useful – a moment's reflection will suggest that it would be difficult to say much if one were not allowed to be vague!

I think the only way to formally understand the prevalence of vague terms is in a model with a different kind of bounded rationality than is considered in this book or in the literature.[5] There are at least three possibilities, which I give in increasing order of ambitiousness. First, vagueness may be easier than precision, for the speaker, listener, or both. For the speaker, deciding which precise term to use may be harder than being vague. For the listener, information which is too specific may require more effort to analyze. With vague language, perhaps one can communicate the "big picture" more easily. This requires a different model of information processing than any I know of.

A more difficult approach would be to derive vagueness from unforeseen contingencies. If the speaker does not know all the possible situations where the listener would use the conveyed information, it may be optimal to be vague. For example, contracts often use vague terms such as "taking appropriate care" or "with all due speed" instead of specifying precisely what each party should do. If agents fear that circumstances may arise that they have not yet imagined, then they may avoid precision to retain flexibility. Hence the optimal contract may require the

---

[5] I give a fuller defense of this position in Lipman (1999).

parties to respond to unexpected circumstances "appropriately," with the hope that the meaning of this word will be sufficiently clear *ex post*.[6] Given the difficulty of modeling unforeseen contingencies (see Dekel, Lipman and Rustichini, 1998, for a survey), this approach is surely not easy.

Finally, I turn to a still more ambitious approach. To motivate it, consider one seemingly obvious reason why vague terms are useful: the speaker might not observe the height of an individual well enough to be sure how to classify him precisely. If we modify the example to include this, however, 1 would have subjective beliefs about Mr. X's height and the efficient language would partition the set of such probability distributions. Hence this objection simply shifts the issue: why don't we have a precise language for describing such distributions?

An obvious reply is that real people do not form precise subjective beliefs. In other words, it is not that people have a precise view of the world but communicate it vaguely; instead, they have a vague view of the world.[7] I know of no model which formalizes this. For example, it is not enough to replace probability distributions with nonadditive probabilities. If agents have nonadditive probabilities, then it is surely optimal to partition the set of such beliefs precisely.

To sum up, I think this book gives some intriguing starts on some important problems. As such, it opens the door to some exciting and difficult research.

REFERENCES

Dekel, E., B. Lipman and A. Rustichini (1998) "Recent Developments in Modeling Unforeseen Contingencies," *European Economic Review*, 42, 523–42

---

[6] This idea is very similar to the Grossman–Hart–Moore approach to incomplete contracts. See Hart (1995) on this approach and Dekel, Lipman and Rustichini (1998) for a discussion of the connection between it and formal models of unforeseen contingencies.

[7] The dividing line between unforeseen contingencies and this kind of vague perception is itself quite vague!

Hart, O. (1995) *Firms, Contracts, and Financial Structure*, Oxford: Clarendon Press

Keefe, R. and P. Smith (1996) *Vagueness: A Reader*, Cambridge, MA: MIT Press

Lipman, B. (1999) "Why is Language Vague?," Working Paper

INDEX

Index